Transductions

TECHNOLOGIES:
STUDIES IN CULTURE & THEORY

Editors: Gary Hall, Middlesex University, and Chris Hables Gray, University of Great Falls

CONSULTANT EDITORS

Parveen Adams, Keith Ansell Pearson, Jim Falk, Steve Graham, Donna Haraway, Deborah Heath, Manuel DeLanda, Paul Patton, Constance Penley, Kevin Robins, Avital Ronell, Andrew Ross, Allucquere Rosanne Stone

Technologies is a series dedicated to publishing innovative and provocative work on both 'new' and 'established' technologies: their history, contemporary issues and future frontiers. Bringing together theorists and practitioners in cultural studies, critical theory and Continental philosophy, the series will explore areas as diverse as cyberspace, the city, cybernetics, nanotechnology, the cosmos, AI, prosthetics, genetics and other medical advances, as well as specific technologies such as the gun, telephone, Internet and digital TV.

BOOKS IN THE SERIES

The Architecture of the Visible: Technology and Urban Visual Culture	Graham MacPhee
The Cyborg Experiments: The Extensions of the Body in the Media Age	Joanna Zylinska (ed.)
Transductions: Bodies and Machines at Speed	Adrian Mackenzie

TRANSDUCTIONS

Bodies and Machines at Speed

ADRIAN MACKENZIE

continuum
LONDON • NEW YORK

CONTINUUM

The Tower Building, 11 York Road, London, SE1 7NX

370 Lexington Avenue, New York, NY 10017-6503

www.continuumbooks.com

First published 2002

British Library Cataloguing-in-Publication Data

A catalogue record for this book is available from the British Library.

    ISBN 0-8264-5883-1 (hardback)

    ISBN 0-8264-5884-X (paperback)

Library of Congress Cataloging-in-Publication Data

Mackenzie, Adrian, 1962–

      Transductions : bodies and machines at speed / Adrian Mackenzie.

      p. cm.

    Includes bibliographical references and index.

    ISBN 0-8264-5883-1 — ISBN 0-8264-5884-X (pbk.)

      1. Technology—Social aspects. 2. Technological forecasting. I. Title.

T14.5 .M315 2002

303.48′3—dc21   2001047375

Typeset by CentraServe Ltd, Saffron Walden, Essex

Printed and bound in Great Britain by

MPG Books Ltd, Bodmin, Cornwall

Contents

Acknowledgements   vii

Preface   ix

Introduction   1

Chapter 1.   Radical contingency and the materializations
             of technology   29

Chapter 2.   From stone to radiation: the depth and speed
             of technical embodiments   57

Chapter 3.   The technicity of time: 1.00 oscillation/second
             to 9,192,631,770 Hz   87

Chapter 4.   Infrastructure and individuation: speed and
             delay in Stelarc's *Ping Body*   116

Chapter 5.   Losing time at the PlayStation: real time and
             the 'whatever' body   145

Chapter 6.   Life, collectives and the pre-vital technicity
             of biotechnology   171

Conclusion   205

References   219

Index   229

Acknowledgements

This book owes a lot to other people. Celia Roberts encouraged me at every step in its writing, and made many generous contributions, both theoretical and practical. A group of people read or responded to chapters at various times. Amir Ahmadi read the opening chapters in detail and joked that it could be a good book. Linnell Secombe and Cathy Waldby read chapters carefully, and brought to light problems I had not recognized. At an early stage, Richard Beardsworth pointed out new directions. In the background, Paul Patton has been a firm support for a long time. I would like to acknowledge the support of the Australian Research Council, under whose postdoctoral fellowship scheme I had the opportunity to write most of the book. Finally, the series editor, Gary Hall, read the book in its entirety and tactfully prompted me to improve it in countless ways.

Preface

Now the rockets fire. You feel the big push. Your moon
ship gets away fast. THEN. . . .

Mae and Ina Freeman, 1962

Getting off the ground, let alone into space, was represented as a
technological sublime, a crowning achievement of *Homo faber*, during
the twentieth century. A 1960s picture storybook for beginning readers
called *You Will Go to the Moon* (Freeman and Freeman, 1962) promised
space travel to Anglo-American — and Australian — boys. At the start
of the book, a neatly dressed child sits on a stool at his living-room
window, observing the moon through a telescope. His father and
mother (the only woman to be seen in the book, and only briefly)
relax on the couch. A train set and model plane lie on the floor, the
already half-abandoned toys of an earlier modernity. By the end of the
book, in a surprisingly rapid sequence of events, the boy has rocketed
to the moon, leaving his nuclear family behind. Suited-up, he and a
'rocket man' stand on a lunar mountain looking towards another
frontier, Mars. Below them lies the 'moon house . . . where you will
live on the moon' (p. 61).

The literal promise of that storybook has not been kept. I have not
gone to the moon. The sight and sounds of blast-offs, launches, orbits,
and splash-downs have become stock footage, and space junk is piling
up in decaying orbits, if it is not dropping out of the sky altogether. In
fact, the only experience I have of space technology is what I have
seen on television screens and in museum exhibits. The Hall of Space
and Flight at the Powerhouse Museum of Technology in Sydney, for
instance, exhibits a relic of the Cold War space race, a recreated space
module. It stands on a pedestal at one end of the gallery. Steps lead

up to a cylindrical module lined with equipment, empty spacesuits, dials and nozzles. Astronauts lived there for weeks, drifting past each other with clipboards and Omega watches. Even on a Saturday afternoon, this end of the museum is not too crowded. It's a long way from the entrance, and more up-to-date interactive computer displays net most visitors before they get this far. A man, a boy and a small girl enter the hall and veer towards the spacecraft. They climb the steps, the boy in front, the girl lagging behind. The boy reaches the platform and runs into the module. The man follows. Behind them, the girl stamps her feet on the steps and grouches: 'Boring old space.'

In miniature, that everyday incident encapsulates the core problem addressed by this book. Everyone has heard that technology will make the future better. Everyone has heard that it could make it a lot worse. The problem is to figure out what scope or capacity there is to collectively manœuvre between these two somewhat bloodless abstract possibilities. *You Will Go to the Moon* was not alone in promising speed, new frontiers, and a homogeneous, clean, frictionless masculine culture of enormous rockets, well-sealed spacesuits and lunar colonies. That future, in which machines will overcome all obstacles, was and still is being often rehearsed today. High technology can symbolize the evaporation of political, religious, racial, ethnic, economic and gender differences. 'Now there is not one thing to hold your moon ship back' (p. 39), the boy is told after blast-off. In many areas, such as biotechnology, information and communications, that same rendering of technology pulls enormous crowds.

At the same time, as a response to that kind of technological bombast, 'boring old space' has its limitations. Admittedly, space technology itself has been somewhat of a disappointment. It seemed to lose momentum once the Cold War spark went out of it. So 'boring old space' is an apposite retort to the museum's veneration of an obsolete technological fetish, of a future that was broadcast as imminent and important. Outside the museum halls, in the humanities, the typical response to the technological future has shared the girl's view: in general, technology is irrelevant, boring, subsidiary or, worse, hostile to culture, tradition, history and meaning. Critical thought has learnt to discount monotonous ideologically loaded talk of bright technological futures partly because of this perception of 'technology' as a cover for troubling differences and conflicts within human collectives. Well beyond humanity's anxieties, anyone who has seen Arnold Schwarzenegger in *Terminator* knows that technology also can symbolize a devastated future. Machines could make us do what we don't want to do. They could foreclose the future. Or more soberly, as historian Leo Marx writes, technology is 'a bloodless abstraction that represents no particular person or thing, no specific skill, vocation or other institution' (Marx, 1999, 144–5).

The point here is that these responses – either straining eagerly towards an impossible technological fantasy, or resisting or ignoring it – gloss over that zone of our collective which is actively engaged, or as I shall be suggesting, *transduced* in technical mediations. From this angle, the problem is how to engage with a collective's embodiment of technical mediations without repudiating, or over-identifying with, technology. The concept of *transduction* is meant to develop some theoretical traction in this still somewhat obscure but richly diverse middle domain. It tries to show how technologies are both difficult to access in terms of subjects and societies, yet indissociably folded through collectives and cultures. That middle domain is lived collectively and it is eventful. Technology is within us, whether we thought we were getting into *or* out of it. To remain with the example of the space race: in its wake military/scientific/commercial satellites crisscross the sky day and night, and, even more to the point, an intricate communication infrastructure is in place. When *You Will Go to the Moon* says on the last page 'then YOU will *see*' (p. 63), it was adopting a slightly more situated outlook on the future of high technology: the space race was played out as a media spectacle, and it spun off media events like a comet's tail.

More generally, habitual desensitization to the stale and over-used word 'technology' does not seem to inhibit the growth and webbing together of technical infrastructures, even if they do not have the high profile of rockets, space shuttles and space stations. No doubt, numbed by the bloodless abstraction 'technology' and tired of its ideological baggage, there may be some reluctance or even incapacity within the humanities to fiddle around with the specific patterns and events associated with the multiple intersection of technologies. Yet, justifiably spurning utopian or dystopian talk of 'technology' as a universal, critical thinkers are landed with a problem: on what ground can anyone engage critically with specific technologies? This is a double bind. Learning to read *You Will Go to the Moon*, for instance, means learning to hear the subjectifying voice addressed to certain groups in the late twentieth century via the symbols of high technology, and at the same time, developing a feel for what always accompanied it: a complicated, messy, fluctuating tangle of technical mediations and collectives involving specific bodies and times. In that sense, *Transductions* is an attempt at learning to read again.

Introduction

A double bind applies to contemporary collectives. Against, through or with technologies, they test their capacities to make sense of their own situation. Yet they are burdened by an overloaded, stale abstraction, 'technology', which overrides and attenuates a more polyphonic voicing of technologies. This book responds to the confusion and ambivalence which flows from that double bind. It is not the first attempt to deal with the topic, and it stands in the shadow, perhaps too much so, of the diverse work of other thinkers such as Gilbert Simondon, Donna Haraway, Bruno Latour, Bernard Stiegler and Martin Heidegger.

There are many ways into this domain. *Transductions* takes a fairly theoretical path, orienting its engagement with contemporary technologies by two problematic reference points: corporeality and temporality. There is no absolute justification for the privilege I am according to bodies and time as points of reference. These areas stand out at the moment for a loose cluster of reasons. Broadly speaking, it is in relation to bodies and time that modern technology effects its most intimate synthesis with cultures. For some time now, living bodies have been seen as 'under attack', or about to be liberated (from nature, and ultimately from death) by communication, information and biomedical technologies. Similarly, the perception has been that time itself is being colonized by speed. Acceleration and instantaneity seem to intolerably compress time. More abstractly, in the general field of recent philosophical thought from which this book takes some important cues, body and time mark limits for thought. To think 'the body' or to think 'time' is to run up against the limits of thinking. Body and time challenge the prerogatives of thought. These two limit cases have undoubtedly triggered important theoretical innovations which have not been clearly articulated together with technology. Our collectives are exposed to an *ongoing* technological dynamism which can perhaps begin to be thought in its nexus with bodies and time.

This approach turns away steeply from any generalized assertion of the effects of new technologies on society. That assertion coasts along on a separation between technology and society which I seek to undermine, if not collapse, at a fairly low level. It will be strikingly apparent from the outset that many specific political, ethical, economic, cultural or social problems efflorescing around technology are under-represented in this book. Recent and oncoming technological changes pose globally large problems, but the focus here is more on the conditions of intelligibility of technological practices in their specificity, on the interleaving of technoscience, capital and cultures. In simple terms, until we have better ways of articulating the technical constitution of our collectives, any general estimation of technological impact remains premature and perhaps mistaken. The volatile essence, the mutable, divergent and eventful character of technologies within our collectives eludes classification as merely 'technical' or 'social'. The interplay between what counts as social/cultural and what counts as technical is far more convoluted than most existing accounts admit.

TECHNOLOGIES, BODIES AND TIMES

Certain anxieties and expectations are insistently raised by the flux of technological change. They are mentioned constantly in the mass media, they figure in many different narratives, and occupy prominent positions in public debates. They revolve around the question of what possible points of orientation we can have when almost every foundation – place, tradition, self, other, body, life, death, culture, nation, history – seems about to be altered, if not utterly transformed, by technology. One question is how such anxieties arise: what drives them? I want to suspend that question, and slow down a little to ask: can we think through the conditions under which such apprehensions of technology arise? This might seem like an obtuse question. However, it is a way of holding onto the double bind mentioned above.

On the one hand, we are already diversely technological. On the other hand, 'technology' has become a problem-fetish for us. The concept of transduction can help frame this question of the 'thinkability' of technology. It both highlights a margin of contingency associated with technological objects and practices within collectives, and clarifies some of the difficulties collectives have in making sense of technical practices. This is not to simply say that technologies can have different meanings, different uses or different effects within any given social context. Rather, this margin of contingency or indeterminacy participates in the constitution of collectives. It entails openness to future and past. The following chapters explore some corporeal and temporal implications of this indeterminacy and openness.

As a way of introducing the concept of transduction, I approach it first of all from the perspective of the deconstructive quasi-concept of 'originary technicity'. Admittedly, this is not an easy way to come up to speed on the topic. However, originary technicity does help remind us that the problem of thinking about technology is also a problem of thinking about time, corporeality and, indeed, thinking about thinking. At a fairly deep level, it unsettles the certainty that we know what technology is in principle. I am assuming that the deconstructive logic of the supplement, with which originary technicity has strong affinities, is familiar enough to clear the path a little here. For the purposes of my general argument, the discussion of originary technicity is in fact mainly a ground-clearing move. Second, I sketch how the margin of indeterminacy just mentioned can be understood in terms of the notion of 'technicity' developed by Gilbert Simondon. Finally, I explore the link between technicity and transduction. We will see that technical objects exist transductively. The agenda here, and in the chapters that follow, is straightforward. Until we can think of technical objects, machines, ensembles in their own terms, then their role in constituting who or what we are remains shrouded. The intelligibility of our own anxieties about technology is entwined with the way we think about technology.

ORIGINARY TECHNICITY AND THE
MEANING OF 'TECHNOLOGY'

It is tempting to use the word 'technology' to mean the whole toolkit of technical artefacts, diverse and innumerable as they may be. However, I am not sure it is possible to do that (and this provides the first cue for the concept of transduction). This difficulty was signalled strongly in the work of Martin Heidegger (Heidegger, 1954, 1977). When he grafted technology onto the question of Being, he exemplified perhaps the most exorbitant use of the term 'technology' in philosophy to date (the German word *der Technik* does not map directly onto 'technology'; Heidegger's response to modern technology will be discussed in Chapter 5). Roughly paraphrased, his work on technology asks: how can we deal with the fact that technology today displays itself everywhere as a constantly shifting, open-ended and groundless ordering of everything that exists, and yet we find it almost impossible to think about how we are collectively involved in that ordering, except in terms of an increasingly untenable anthropocentrism which elevates us, as 'the human', to the summit of all things (Heidegger, 1977)? Put differently, when IBM Corporation's supercomputer *Deep Blue* beat the reigning human world chess champion, Gary Kasparov, in 1997, what was our reaction? To say that humanity is still superior in the contest of intelligence because it built the technology of supercomputing? Heidegger's questioning of technology addressed the profound incoherence of such a response. He sought to think through the experience of being suspended between growing orderability and an as yet indecipherable historically specific involvement with technology.

There is no simple way out of that experience. It is another version of the double bind. Could we then move up a level and, speaking more reflexively, say the technology is a set of artefactual, corporeal and semiotic realities, folded into each other through the broadly delimited experience of an historical situation (e.g. progress, utopia, dystopia, etc.) as modern? From this more sophisticated perspective, the word 'technology' would be treated as a historically situated discursive entity, dating from some time in the early nineteenth century or a little earlier (Mitcham, 1994, 130–31). Any reference to technology would then inescapably filter through this complex historical siting. The unstable affective charge that the term technology carries today, whether it be Silicon Valley's enthusiasm, Wall Street's panicky glee, consumer resistance to genetically modified (GM) foods, or the girl's boredom with space as the final technological frontier, would thus connote the fragmentation of that historical experience. 'Technology' could then be treated as a discursive reality generated by the historical processes of modernity.

The problem with this move is that, for better or worse, technology is more than a manner of speaking or launching new commodities or political programmes. It strongly resists reduction to discourse and signification. Rather, it tends to condition them. This is not to say that technology, or some aspect of technology, is outside discourse. Rather it is to say that we can think, signify, make sense and represent who we are in part only because of technology. Obliquely departing from Heidegger's work, recent work in continental philosophy has homed in on this point. Drawing primarily on deconstructive approaches, a cluster of French and British theorists including Jacques Derrida, Jean-François Lyotard, Bernard Stiegler, David Wills, Geoffrey Bennington, Richard Beardsworth and Simon Critchley have referred to a collective, constitutive human exposure to something tentatively called 'originary technicity'. This difficult and awkward technical term shies away from any simple substantive definition. Keith Ansell Pearson outlines it in these terms:

> Current continental philosophy contends that the human is necessarily bound up with an originary technicity: technology is a constitutive prosthesis of the human animal, a dangerous supplement that enjoys an originary status. (1997, 123)

He refers to Derrida's work, which from time to time invokes the notion of originary technicity. Since it directly links originary technicity to corporeality, the following from Derrida is relevant:

> The natural, originary body does not exist: technology has not simply added itself, from outside or after the fact, as a foreign body. Certainly, this foreign or dangerous supplement is 'originarily' at

work and in place in the supposedly ideal interiority of the 'body and soul'. (Derrida, 1993, 15)

One tack we could take on this quasi-concept of originary technicity is to say that it concerns the status of the body as a body. It may not be possible to think of a body *as such* because bodies are already technical and therefore in some sense not self-identical or self-contained. In *Specters of Marx*, the point is developed further:

> [W]hatever is not the body but belongs to it, comes back to it: prosthesis and delegation, repetition, differance. . . . To protect its life, to constitute itself as unique living ego, to relate, as the same, to itself, it is necessarily led to welcome the other within (so many figures of death: differance of the technical apparatus, iterability, non-uniqueness) (Derrida, 1994, 141)

If there is no non-technical body as such, what we call a living body will have and has already had to admit 'others within'. A body might have to be approached under the general deconstructive rubric of iterability. Distinguishing the dynamics of originary technicity from the dynamics of technology as they are usually understood, Geoffrey Bennington observes, The dynamic of technicity will thus be the dynamic of the prosthetic – and thereby the human as non-proper supplementarity – in general (Bennington, 1996, 181). (It would be possible to cite formulations from Richard Beardsworth and also Simon Critchley that make the same point (Beardsworth, 1995, 1998; Critchley, 1999)). The mutability and eventfulness of technology should be neither approached as autonomous agent nor found entirely wanting in dynamism compared to the living. Most explicitly of all, Bernard Stiegler has expansively argued in his multi-volume work, *La technique et le temps* (Technology and time), that the notion of originary technicity draws out certain implications of deconstructive thought that have been more or less overshadowed in its literary and philosophical reception (Stiegler, 1994, 1998). Rather than treating technology as textual, all of these deconstructive treatments suggest that textuality, discourse, meaning and life more generally, is already technical without being, for all that, technologically determined. The line they mark between the technology as discursive entity and technology as global-izing ordering of communication and production is complicated, unstable and divisible.

In terms that markedly stray from this strongly Western European, deconstructive approach, I think that the feminist historian and theorist of science and technology Donna Haraway points to something similar when she writes, 'I define corporealization as the interactions of humans and nonhumans in the distributed, heterogeneous work pro-cesses of technoscience' (1997, 141) or 'the body is simultaneously a historical, natural, technical, discursive and material entity' (p. 209). She could be understood as saying that what we take to be a body is not only inseparable from technologies (or 'technoscience'), but dynamically engendered in the interplay of disparate actants. The important point, highlighted in the second citation, is that bodies figure as sites of complication, intersection and heterogeneous collective processes. The sociologist of science and technology, Bruno Latour, also strikes out in this direction at times, when he writes, for instance, that 'there is no sense in which humans may be said to exist as humans without entering into commerce with what authorizes and enables them to exist (i.e. to act)' (Latour, 1994a, 45–6); or, 'even the shape of humans, our very body, is composed in large part of sociotechnical negotiations and artefacts' (p. 64). Again, as in the deconstructive accounts, stress falls on living bodies as the domain in which what it is to be human encounters something other, and finds itself in 'com-merce' with an enabling other.

It is worth noting how many of these formulations directly and inextricably associate living bodies with technical action. The associ-ation is more than an external linkage between bodies and technical artefacts. The adjective 'originary', as some of the quotes just given indicate, is one way to describe something more unnerving and unlocatable than merely strapping on, implanting or even injecting gadgets into living bodies. By now, 'originary' has become familiar shorthand for the deconstructive logic of the supplement. The logic of the supplement describes all those situations in which what was thought to be merely added on to something more primary turns out to be irreversibly and inextricably presupposed in the constitution of what it is said to be added on to. Derrida writes in *Speech and Phenomena*: 'the strange structure of the supplement . . .: by delayed reaction, a possibility produces that which it is said to be added on' (Derrida, 1973, 89). If, as an earlier quote from Derrida indicates, technology can be seen as a 'dangerous supplement', we may be justified in saying that the apparent 'adding on' of technology to living bodies has a complex temporal structure. It may be necessary to think about technicity in relation to time. This will be the other direction in which originary technicity takes us: towards an engagement with technology as temporal, or more correctly, as temporalizing. To speak of the inextricability of bodies with technology is also and always to speak of time.

WHAT DOES TECHNOLOGY SUPPLEMENT?

What is supplemented by technicity (a term whose specific meaning will soon be discussed)? In nearly all the formulations I have quoted, it is 'the human animal', 'humans', or 'the human'. More generally, it is 'life' or 'the living.' Now according to the logic of the supplement, if technicity supplements the living, the non-technical does not simply precede the technical. Conversely, technologies are not simply added on to cultures, for instance. So for instance, an essential human capacity to use tools cannot pre-exist the development of those tools (see Chapter 2). Originary technicity implies that the non-technical entity, the 'who' or the human, also has need of the 'what', the technical supplement, in order to become who she or he is. The point would not be to posit primacy for either the technical or the non-technical, but to see how the secondary position of one term (technicity) allows the other term to both be thought and remain in some sense unthought.

One way into this difficult terrain has been opened up by Bernard Stiegler. The deep interlacing of technology and time forms the principal focus of his recent work (*La technique et le temps. 1. La faute d'Épiméthée* (The fault of Epimetheus); *La technique et le temps. 2. La désorientation* (Disorientation)). Along with philosopher Gilbert Simondon's work (to which we will turn in just a moment), Stiegler's work offers one of the most elaborate and wide-ranging accounts of originary technicity currently available. (The major critical responses to Stiegler's work available to date in English are Bennington 1996; Beardsworth, 1995, 1997, 1998.) Stiegler brings the delayed reaction implied by the logic of the supplement to the forefront. In a strongly Heideggerian vein, he argues that the relation between technology and culture is a kind of historically materialized 'temporalization'. He writes:

> Technology evolves *more quickly* than culture. More accurately, the temporal relation between the two is a tension in which there is both advance and delay, a tension characteristic of the drawing-out [*Erstreckung*] which makes up any process of temporalization. (Stiegler, 1993, 43)

The term 'temporalization' is Heidegger's (1962, 386–7). It would be difficult to explore it in detail here. Conceptually, it could be seen as a precursor to the logic of the supplement. The important point is that, for Heidegger, time is not an entity or substance which would simply have a past, present and future as its attributes. Nor does it designate the unstable appearances of an underlying reality. Rather, temporality is an openness or disjunction affecting every level of what exists. Temporality temporalizes itself variously. For entities who think about it, it proves particularly troublesome. Existentially, it means that we do not exist simply in ourselves, but hold ourselves open to a future that we cannot fully appropriate (e.g. I will not experience my own death), and find ourselves unaccountably affected by a past that precedes us. By analogy, Stiegler is arguing that the relation between culture and technology also in some sense temporalizes, or makes possible, a relation to future, past and present.

There are many questions that such a brief foray into technology as temporalization cannot answer. Strategically, Stiegler advances the idea that technologies temporalize rather than flatten time out. Apart from all the conceptual complications that Stiegler's work at times entails, this is a novel affirmative thesis which warrants serious consideration. For the moment, all I want to suggest is that approaching time and technology from the angle of original technicity has a better chance of negotiating the double bind than some other more conventional response approaches. For instance, technology is often viewed apprehensively as an unstoppable juggernaut. From Stiegler's perspective, the technical runs ahead of culture, but it is not alone. It enlists humans to power its instantiation. As a supplement, it is not autonomous or intrinsically dynamic. This very complicated point awaits fuller discussion (see Chapters 4 and 5). The main idea is that when we think about originary technicity, we can expect to find a complicated interlacing of anticipation and delays. There is instability and movement at the joint between technology and culture, but this is not because either is an autonomous agent.

ELEMENTARY TECHNICITY AND DELOCALIZATION

Originary technicity brings the deconstructive logic of the supplement into play around technology. That means that when we try to decide whether humans and technologies are entwined corporeally and temporally, we cannot ground our judgements in a radically non-technical domain. Clearly, this must have consequences for the question we are tracking here. I asked: can we think through the conditions under which apprehensions (fearful, eager) of technology arise? A response informed by the quasi-concept of originary technicity would say that we can render those conditions intelligible in ways which are already marked by something technical. Thought, in other words, has its technicity. The logic of the supplement takes us a long way from ideas of technology as material artefact or ideological abstraction. It might help to explain how such abstractions gain traction. Meanwhile, the other term, 'technicity', has been put aside. It can be brought forward to show how a margin of indeterminacy is associated with technology that neither belongs solely to human life nor belongs to some intrinsic dynamism of technology.

As it appears in recent French thought, the notion of technicity perhaps stems most directly from the relatively little known yet startlingly fresh and relevant work of the philosopher Gilbert Simondon. Whereas originary technicity stresses a dehiscence in concepts of the (human) subject, technicity, as developed by Simondon, emphasizes something similar in technical objects. Simondon was a student of Georges Canguilhem, and has published several books and many articles on technicity, technology, 'individuation', affect and collective (Simondon, 1989a, b, 1992, 1995; Combes, 1999; Hottois, 1993;

Dumouchel, 1995). When current deconstructive thought invokes the term 'originary technicity', it is hard to not hear first of all an echo of the concept of technicity Simondon developed in the 1950s and 1960s, if only because the term is not often found elsewhere. Simondon initiated an important shift in perspective on technology through the notion of technicity. In drastically summarized anticipation of the consequences of his approach, we could say that the concept of technicity refers to a side of collectives which is not fully lived, represented or symbolized, yet which remains fundamental to their grounding, their situation and the constitution of their limits. Technicity interlaces geographic, ecological, energetic, economic and historical dimensions without being reducible to any of them.[1]

In Simondon's first book, *Du Mode d'existence des objets techniques* (The mode of existence of technical objects) (1958), the term 'technicité' occurs repeatedly. The book seeks to redress a misguided opposition between culture and technology. Such an opposition may have long existed, but was less hampering when the technical objects in question were more discrete and less extensive and potent than recent or current technologies. Broadly speaking, Simondon argues that a misapprehension of the way in which technical objects exist prevents us from seeing their part in the constitution of human collectives, or in 'the human'. The book also deals with the problem that large-scale technical *ensembles* (such as information, communication or transport infrastructures, biotechnological interventions, etc.) pose for thought, representation and collective life. These ensembles are difficult to represent as such because of their sprawling, distributed and often quasi-invisible existence. Simondon's response to both the opposition between culture and technology, and the problem of representing technical ensembles relied on the concept of technicity. Technicity plays a major role in re-evaluating what a technical object is, whether it be a tool, a machine or multi-system ensembles or infrastructures, and thereby opens the possibility of a conceiving collective life somewhat differently.

CONCRETIZATION AS GENESIS

What is this promising concept of technicity? Unfortunately, there is a major obstacle to answering the question quickly and neatly, and this problem arises from a core difficulty in recognizing and figuring technology within the life of a collective. The technicity of something like a handtool can be provisionally isolated from its context. A handtool is, in Simondon's terms, a *technical element*. The technicity of 'a technical element' might materialize, for instance, in the different zones of hardness and flexibility combined in a blade that cuts well. That combination reflects a thoroughly localized assemblage of practices. They are so localized as to be given proper names: 'Toledo steel' or 'Murano glass'. Paradoxically, such names reflect the fact that these low-technology artefacts are more de-localizable than the extensive networks of technical mediation that characterize contemporary technology. The technicity of a technical element is more mobile or detachable than the technicity of an ensemble which is always in situ. So, ironically, despite the techno-hype, the most up-to-date, high-speed technical mediations are in Simondon's terms perhaps the least mobile, the most heavily constrained and weighed down by their context. A mobile phone or wireless appliance could be understood from this perspective as a massively encumbered object. Its physical portability and miniaturization comes at the cost of an increased ramification and layering of communication infrastructure. Because they can be detached and mobilized in different contexts, Simondon says that 'it is thus in the elements that technicity exists in the most pure way' (1989a, 73). By contrast, it is not possible to directly distill the technicity of a large-scale technical ensemble. Even the technicity of a machine (such as an engine) cannot be isolated from the 'associated milieu' which it inhabits. That milieu (which includes flows of air, lubricants and fuel, for instance) conditions and is conditioned by the working of the engine. In the case of an ensemble, we would need to investigate how the technicity of sub-ensembles enter into commutation, and mutually condition each other.

As its technicity is heightened, an element becomes more stable, or detachable from its context. It becomes mobile, and its effects become more iterable. It is significant that even here in the case of 'pure' technicity, the technicity of an element still derives from an ensemble. The technical element carries with it something acquired in a situated, grounded ensemble. For a long time, a steel blade made in Toledo had a reputation for hardness, flexibility and durability that stemmed from a combination of the local charcoal, the chemical composition of the water, and the forging techniques used. Often the technicity of a technical element (something which enters into the composition of a technical object) reflects a complicated and even globally extended technical ensemble in its own right. Semiconductor chips exhibit that kind of technicity. Also, as we will see in a later chapter, the technicity of a simple technical element such as a stone hand-axe implies intensive corporeal organization. For an isolated technical element, technicity refers to the degree of *concretization* which the intersection of these diverse realities embodies. Note that we are already in the domain of a transductive process here: technical elements possess a degree of concretization because they encapsulate a singular combination acquired in an ensemble. The hallmark of a transductive process is the intersection and knotting together of diverse realities. (The next chapter will analyse a brick from that perspective.)

Technical elements, such as a spring, a wheel, a cutting edge, a switch, a logic gate or a monoclonal antibody, embody a capacity to produce or undergo certain specific effects. The technicity of an element is heightened or diminished according to the relative independence it displays in relation to variations in context. It consists in the 'capacity of an element to produce or to undergo an effect in a determined fashion' (Simondon, 1989a, 72–3). It is that 'quality of an element by which what has been acquired in a technical ensemble expresses and conserves itself in being transported to a new period' (73). Technical objects actualize or instantiate their technicity in various degrees of abstraction or concreteness.

THE PROBLEM OF ENSEMBLES

How does this relate to the question of the intelligibility of responses to technology? We have already glimpsed the implications of the quasi-notion of originary technicity. It implies a contamination of thinking about technology by technicity. I am now suggesting that Simondon's problematization of the technicity of ensembles permits a more fine-grained analysis of why it might be difficult to orient ourselves in relation to technology. Technicity pertains to a kind of iterability associated with the *technical elements* and derived from a singular, site-specific conjunction of different milieus. Technicity can be found within different contexts broadly ranging between small sets of tools to ensembles composed of many sub-ensembles. The technicity of a hand-tool implies something different to the technicity of a telecom-munications network or a semi-autonomous machine. Even if, as we will see in a later chapter, the technicity of a hand-axe cannot really be ranked *lower* than that of a supercomputer, it is the technicity of the ensemble that is particularly problematic.

As a first step in his problematization of technical ensembles, Simondon argues that technical objects need to be understood in terms of their genesis, rather than as stable objects. This is not a suggestion that we should simply take an historical perspective on technological development. Rather, he is emphasizing that the very mode of existence of machines and ensembles implies sometimes divergent tendencies which provisionally stabilize in specific technical objects. The essentially genetic existence of technical objects springs from the variable consistency or concretization embodied by different specific technical objects. A technical object lies somewhere between a tran-sient, unstable event and a durable, heavily reproduced structure. Its degree of 'concretization', to use Simondon's terms, is the technicity of a technology.[2] In these terms, a high technology can possess a low technicity. The genesis of a stable entity is implicitly a transductive process. Simondon describes technicity as 'a unity of becoming' (1989a, 20), and as a network of relations:

> Technicity is a mode of being only able to fully and permanently exist as a temporal, as well as spatial, network. Temporal reticula-tion consists of resumptions of the object in which it is reactualized, renovated, repeated under the very conditions of its initial fabrica-tion. Spatial reticulation consists in the fact that technicity cannot be contained in a single object. An object is only technical if it occurs in relation with other objects, in a network where it takes on the meaning of a keypoint [point-clef]; in itself and as an object,

it only possesses virtual characters of technicity which actualize themselves in active relation to the ensemble of a system. (1958, 325)

Once we think about technical objects as existing genetically, technicity exists as a network of references or relays. Even if a technical element exists discretely, its technicity is deployed in relation to other elements and gestures, to other practices and institutions. Pure technicity is very elusive, because technicity endures or persists through dispersed, even discontinuous, repetitions across clusters of technical elements in interaction.

Having understood technicity in genetic terms, the reason why ensembles pose a problem for thought becomes visible. They are composed of the technicity of their technical elements. Technical ensembles assemble and organize the technicity of elements, not by forming matter. To see technical action as assembling technicities involves a specific and nuanced critique of the hylomorphic or matter-form schema that has regulated most understandings of technical action and many philosophical notions of what a body is since Plato. (This will be discussed in the next chapter; see Simondon, 1989a, 74.) However, it also raises the problem of how to think about the technicity of the ensemble. If the technicity of a technical element is already temporally and spatially reticulated, what about technical objects such as machines and ensembles? There is an almost bewildering topological and temporal complexity here: the technicity of the ensemble is constituted from the spatially and temporally reticulated technicities of its elements.

TRANSDUCTION AND COLLECTIVE INDIVIDUATION

Handling this complexity requires a final shift in the level of analysis from the problem of the ensemble to the concept of transduction. It brings us back into proximity with collective life, something we have not caught sight of for a while. The concept of transduction answers directly the problem of thinking about diverse interactions and resonances between the elementary technicities present in a technical ensemble. At the same time, it also extends to the emergence of resonance and coupling between diverse realities. It occurs around singular points, and it is a process that highlights *metastability* rather than stability in a given context. The problem of the technicity of a technical ensemble feeds directly into the broader problem of thinking through the conditions under which anxieties and expectations about technology arise.

In his later work *L'Individu et sa genèse physico-biologique* (The individual and its physico-biological genesis), Simondon provides a kind of definition of transduction:

> This term [transduction] denotes a process – be it physical, biological, mental or social – in which an activity gradually sets itself in motion, propagating within a given domain, by basing this propagation on a structuration carried out in different zones of the domain: each region of the constituted structure serves as a constituting principle for the following one, so much so that a modification progressively extends itself at the same time as this structuring operation. . . . The transductive operation is an individuation in progress; it can physically occur most simply in the form of progressive iteration. However, in more complex domains, such as the domains of vital metastability or psychic problematics, it can move forward with a constantly variable step, and expand in a heterogeneous field. (Simondon, 1995, 30–31)

Note the continuity between this definition and that of technicity. Both concentrate on ontogenesis rather than ontology. That is, technicity and transduction account for how things become what they are rather than what they are. Technicity is one important kind of transduction: that which pertains to technical objects. A technical element such as a blade, a spring, a switch or a cultivated seed resolves a divergent set of constraints within a given domain. It represents a certain degree of compatibility between them. Again, a machine embodies a technicity that pertains to a collection of elements located at the intersection of different milieus. Finally, an ensemble possesses an even more distributed kind of technicity.[3]

In *L'Individu et sa genèse physico-biologique*, first published in 1964, Simondon generalized the term 'transduction' to name any process (physical, biological, social, psychic or technical) in which metastability emerges. His interest in *ontogenesis* (that is, on how something comes to be) rather than *ontology* (that is, on what something is) stems from a mode of thought focused on a unity of becoming rather than a unity of substance. The spectrum of transductions ranges from simple iteration (as in Simondon's paradigmatic example of a physical transduction, the growth of a seed crystal suspended in a liquid) to constantly varying rhythms oscillating in a field structured by differences and repetitions (as for instance, in affect and thought). Transduction arises from the non-simultaneity or metastability of a domain, that is, in the fact that it is not fully simultaneous or coincident with itself. Boundaries, singularities and differences underlie transductions.

This means that living things can also be understood transductively. (We should note first of all that transduction has specific meanings in recent biology. As we will see in the last chapter, in the 1950s molecular biology began to speak of 'transduction'. Cell biology has also developed a specific meaning for the term 'signal transduction'. In molecular biology, it named a specific event in which a virus carries new genetic material over into the DNA of bacteria. Viral transduction prefigures the forms of genetic manipulation currently under intense development.) Non-living individuation, while transductive, always occurs on the surfaces or boundary between the individuating entity and its milieu. The planes on which the crystal grows are always on those surfaces of the crystal in contact with a liquid. Life is transductive too, but involves temporal and topological complications. The living encounters information, understood strictly as the unpredictability of forms or signals, as a problem. It resolves the problem through constant temporal and spatial restructuring of itself and its milieus. It develops and adapts, it remembers and anticipates. Unlike a crystal, life can individuate (that is, develop in its specificity out of a domain of unresolved tensions and potentials) to a greater or lesser extent by becoming information for itself. It possesses interior milieus. It is as if a crystal could become a medium for its own further growth. Simondon calls that process a 'recurrence of the future on the present' (1989a, 144). The living gives information to itself and, in doing so, individuates itself on the basis of a reserve of pre-individual singularities, or a field of intensities not yet organized in specific forms and functions.[4] Finally, there would be forms of life whose collective individuations includes technicity.

THINKING TRANSDUCTIVELY:
FROM TECHNICITY TO COLLECTIVE

There are still unanswered questions about transduction, and perhaps especially about transduction as a way to think through the double bind between technology as overloaded signifier and technical practices, intimately embodied and situated. The main point is that transduction aids in tracking processes that come into being at the intersection of diverse realities. These diverse realities include corporeal, geographical, economic, conceptual, biopolitical, geopolitical and affective dimensions. They entail a knotting together of commodities, signs, diagrams, stories, practices, concepts, human and non-human bodies, images and places. They entail new capacities, relations and practices whose advent is not always easy to recognize.

A transductive approach promises a more nuanced grasp of how living and non-living processes differentiate and develop. It understands the emergence of a mode of unity without presuming underlying substance or identity. Every transduction is an individuation in process. It is a way something comes to be, an ontogenesis. Importantly, transduction refers not only to a process that occurs in physical, biological or technical ensembles as they individuate. It also occurs in and as thought. Thinking can be understood as an individuation of a thinking subject, not just something that someone who thinks does. To think transductively is to mediate between different orders, to place heterogeneous realities in contact, and to become something different. Correlatively, thought which undertakes to comprehend such processes must itself be transductive if it is to accompany the constitution of individuated entities. A transductive process calls for transductive thought.

Looking back, we can now at least envisage the problem of the double bind between technology as a grand signifier and the diversity of technical practices a bit differently. The first step we took was to say that however we think about technology, there is no way we could purify any other term (human, life, society, politics, subject, ethics, truth) of its technicity. This was a very general point, but it means that anxieties and hopes concerning technology can be read critically, even deconstructively. They themselves are already technically mediated. Our second step was to quickly traverse a finer-grained account of technicity drawn from Simondon's work. Technicity is a concept that diffracts technical objects into a network of temporal and spatial relays. The mode of existence of technical objects is genetic. It involves delocalizing and localizing vectors, and it moves between unstable events and durable structures. In particular, technicity flags the problem of representing the mode of existence of contemporary technical ensembles. The third stage was to say that thinking about technicity opens on to a wide-ranging style of thought focused on individuation, or on the emergence of new capacities in the intersection of diverse domains.

The question I posed was: can we think through the conditions under which technology becomes something to be apprehended fearfully as an alien or hopefully as a saviour? Thinking transductively about this problem entails suspending any prior, separate substantial unity in either technology or the collectives (societies, cultures, civilizations, etc.), and attending to the processes that separate and bind them. This is easy to say, but how does one do it?

SINGULAR EXAMPLES, GENERAL CONCEPTS?

Each of the following chapters traces a thread held in tension by the double bind. Each one takes some idea, figure, perception, experience or affect associated with contemporary technology (speed, power, autonomy, complexity, pleasure etc.), and shows how that double bind between technology as overloaded signifier and concrete practice applies to it. From another angle, the chapters of this book can be seen as plotting a path through recent philosophemes. The chapters take general concepts such as body, materiality, time, community, individuality and life, and engage with them through singularities or key points of technicity in contemporary collectives. Examples such as a brick, a seventeenth-century pendulum clock, a supercomputer, a pre-hominid hand-tool, a performance by the artist Stelarc, an online computer game, a satellite navigation system and a genomic database have been chosen because their singularities strike me as implicated in some way in the more general concepts.

The first chapter discusses corporeality and technicity. The connections between bodies and technologies are currently under intense theoretical and practical scrutiny in many quarters (ranging from critical theory to biotech and pharmaceutical companies), not least because of the increasingly direct biotechnological manipulation of what was held to be in some sense immutable – the limits of life and death – and inalienable – the propriety or 'mineness' of living human bodies. Technical practices often figure as invading living bodies, or liberating subjects from the burdens of embodiment. Representing them in this way risks losing sight of the ways in which our collectives are redistributing and reconfiguring relations between different life-forms and technical apparatuses. Even calling these practices 'biotechnology' obscures the problem to a certain extent. The technicity of these new arrangements resists formulation in terms of existing ideas of what a technology is, and it certainly poses a challenge to many theoretical accounts of embodiment. Drawing on feminist theories of the body, the first chapter develops a transductive account of corporeality. The argument develops out of the more abstract formulations of originary technicity encountered above: what we take to be a body is already in some sense technical, and therefore bodies and technologies couple in ways that are a little more complicated than any simple version of technology as organ extension suggests. By taking into account some of the ways bodies are constitutively and intimately technical, it might be possible to offer an altered account of the propriety of the living body, one which begins to orient itself with respect to fears of a loss of corporeal propriety. The complexity of modern technology is often contrasted with other so-called 'primitive' technical practices. Thinking transductively about embodiment undercuts the self-evidence of that contrast. Such contrasts can obscure the historical-collective existence of technologies.

The second chapter contrasts two historical limit cases: that of stone hand-tools used by proto-hominids and the supercomputers used in nuclear weapons design. These artefacts are not disinterested examples. Hand-tools figure strongly in discourses concerning hominization. They have functioned as a kind of limit-term between nature and culture in philosophical, social-scientific and popular accounts of human origins. In terms of technicity too, they exemplify limits. The hand-tool is a technical element radically detached from the collective in which it was produced. Nuclear weapons are another limit, but this time by virtue of a massive technological ensemble developed under specific economic and political conditions during the second half of the twentieth century. Nuclear weapons systems iconize modern technology becoming autonomous, global, and out of control (Winner, 1977). When stone axes and thermonuclear weapons are compared, the contrast usually loads all power and complexity on to contemporary technology. A transductive account of technologies and bodies again suggests an alternative to this picture. Other kinds of topological and temporal complexity need to be considered. In the case of the hand-tool, those complications concern the process of corporealization: the technicity of a hand-axe, and specifically its capacity to be mobilized apart from the sites of its production, implies patterns of gestures and perceptions which cannot simply be regarded as natural. In the case of the thermonuclear weapon, the complications concern the practices of inscription and calculation which participate in its technicity. The force of the bomb as a technical mediation is difficult to signify except in apocalyptic terms, but this force is intricately interwoven with an organization of traces and inscriptions. Just like the hand-axe, the bomb is embodied in a collective. The apparently relentless historical expansion of modern technology should be re-thought.

The third chapter also relies on a contrast, but this time between a seventeenth-century pendulum clock built by the Dutch scientist Christiann Huygens, and a late twentieth-century clock system, the global positioning system (GPS) deployed by the United States Department of Defense. Again, the choice of example is not arbitrary. Clock-time surfaces in a wide stream of historical and theoretical work on modern technology as responsible for a kind of loss of social or lived time. It figures as the prototype of a global technological imprinting and speeding-up of collective life. It would be futile to deny these effects. Yet viewed transductively, this way of posing the problem moves too quickly to separate technology and collective life. Rather than simply colonizing lived time, clock-time articulates a diverse set of realities on each other. It brings different orders or domains into relation in ways that neither social construction nor technological determinism can grasp. The technicity of the pendulum clock, I will argue, resides at the intersection of geographic, political, military and economic realities. Mutability, metastability and eventfulness are a direct consequence of the transductive processes associated with originary technicity. The notion of transduction has important implications for any experience of *speed*. Recent theories of technical change tend to attribute an absolute value to the speed of contemporary technology. Technological speed is regarded as assaulting subjectivity and life. Prominent theories (e.g. the work of Paul Virilio) speak of a radical break or disjunction in our experience caused by speed, particularly the celerity of media teletechnologies. The consequences are usually presented as catastrophic, apocalyptic or revolutionary. From the standpoint of originary technicity, there is a need to be careful about how we evaluate this experience of speed. As Chapter 3 proposes, drawing critically on Heidegger's work on time and technology, there can be no pure experience of speed, only of differences of speed.

The fourth chapter considers a work by the performance artist Stelarc entitled *Ping Body* and provides one way of moving more slowly around the question of speed. That work, in which a living body transduces a flow of data measuring network response times into gestures and images, shows that any experience of speed already deeply embodies a technical apprehension. There is no pure, non-technical apprehension of speed, nor any non-technical access to time.

ch p. 5

The problem of what kinds of collective emerge from contemporary ensembles of computation and communication surfaces as an altered problem for thought in the next chapter. It takes the example of an online, real-time computer game. Computer games often count as a debased and thoroughly commodified cultural form, at least for critical theories of culture and society. They are seen as impoverishing sociality, rather than as generating new forms of representation. This blanket rejection would blind us to any constitutive role for games and play in the formation of collectives. Treated a little more seriously, these toy artefacts also figure in some of the ways in which the emergence of real-time collectives are coupled to technical infrastructures. The specific temporal dynamics and disjunctions of real-time computation and communication challenge political and cultural theory to invent different ways of conceptualizing collectives.

ch.6   Finally, 'life', a term threaded into the background of many different accounts of technology, can be considered from a transductive perspective. Contemporary developments in biotechnology accentuate the key role and significance of life in technical mediations. Life in its speciated variety, and in its accumulated site-specific history of accidental mutations and variations, has become a primary resource for biopolitical industrialization and commodification. Biotechnological processes, as they have unfolded over the last several decades, have emerged in close relation to informatic and computational technologies. All of this is fairly well known. But a number of problems remain. This chapter addresses two of them. The first is how to formulate the technicity of the ensemble of biotechnical practices. The relations unfolding within that ensemble are not easy to represent, since they entwine strands of heredity, kinship, reproduction, health, property and race, to name a few, with technical systems. In Simondon's terms, the problem can be framed as one of how to think through the technicity of an ensemble whose elements are assembled from non-living and living milieus. As a way into this problem, the chapter examines how genomic information is organized and manipulated in computer databases. The genomic and proteomic databases that store and retrieve sequence data form a sub-ensemble of the complex processes of reassembly involved in biotechnology. The organization and processing of sequence data there can be read as a symptom of a specific kind of technicity. Those manipulations have a singular texture and inscriptive materiality which plays an important part in biotechnology. The second problem is more thematic, and concerns some of the general implications of a transductive approach. The curiously half-living, half-non-living status of biotechnological mediations heightens in important ways the instability of the borderline between life and death. From this angle, biotechnology and what is happening through it might be addressed differently. Rather than seeing biotechnology as threatening the propriety of life, we might see it as making explicit some of the consequences of the technicity of collectives.

The problem addressed in the book can be formulated as a question: how can we acknowledge the powerful global extension of modern technology, with all its dislocating effects of speed, and yet remain responsive to the specific historical layering of collectives composed of humans and non-humans in that event? When Martin Heidegger (1977) talks about reflecting on the essence of technology as a way to face these difficulties, he insists on the necessity of clearing away any contamination of that essence by the merely technical. That insistence, although taken to its limits in Heidegger's work, typifies an important strand in existing responses to technology. When technologies and sociotechnical collectives are read in terms of originary technicity, it may be that this clearing away and decontamination of technology need not be pursued so relentlessly. That could be important in a number of ways. It would mean, for instance, that thought, even philosophical thought, could begin to affirm its own technicity. It would imply that our collectives could begin to articulate their own constitution and limits more explicitly. A reaction that flattens out technical mediations risks moving too quickly. *Transductions* explores some grounds on which we might move a little more slowly. It broaches some ways in which that experience of speed and change indissolubly attached to contemporary technology can also be understood as a collective slowing down.

NOTES

1. Although rarely cited explicitly, it could be argued that Simondon's work also lends impetus in several ways to the accounts of human–non-human agency developed by Bruno Latour, Michel Callon and others under the term 'actor-network' theory (see Chapter 3). These accounts have been influential in recent social studies of science and technology and, more broadly, in cultural and social theory (e.g., Lash, 1999). Very importantly for this book, Simondon maintains that technicity *and* collectives need to be *thought transductively* because they are transductive processes. Any theory that responds to technicity will need to itself be transductive. (Incidentally, Simondon's notion of *transduction* as a way of understanding the temporal and corporeal individuation of living and non-living entities also surfaces in Gilles Deleuze's work (Deleuze, 1994).) Hence transduction designates both a process that lies at the heart of technicity and a mode of thought adapted to

thinking how collectives are involved, as Deleuze puts it, in the 'establishing of communication between disparates' (1994, 246). Transduction names the process that occurs as an entity individuates or precipitates in a field of relations and potentials. Although we could approach Simondon's understanding of transduction from various directions (including physical, biological, psychical and collective processes), the most direct path in this context is via his account of technicity.

2. It should be clear by now that 'technicity' therefore means something radically different from 'technology', understood in the usual English sense of the term as the loosely defined tools, machines and systems mainly associated with human, and to a much lesser extent, non-human actions. The term 'technologic' does play a significant role in Simondon's work, and has done in French and German thought over the last century or so, but it refers mainly to the systematic study of the transformations and correlations that characterize technical objects (Simondon, 1989a, 48; Mitcham, 1994; Siguat, 1994).

3. Even if we cannot readily evaluate the technicity of large-scale technical ensembles, we can say that technicity is in principle the temporal and spatial network of interactions between technical elements which to some degree resolves a specific disparity or incompatibility within a given domain. The notion of transduction opens unusually far-reaching connections between questions of technology, corporeality and time. It could allow both the more or less recent ideological loading of the term 'technology' and the widely varying yet undeniable technicity of human collectives to be held in tension.

We could also approach transduction starting from technical elements known as 'transducers'. The terms 'transduction' and 'transducer' have technical meanings in biology and engineering. In electrical and electronic engineering, transducers convert one form of energy into another. A microphone transduces speech into electrical currents. For the process of transduction to occur, there must be some disparity, discontinuity or mismatch within a domain; two different forms or potentials whose disparity can be modulated. Transduction is a process whereby a disparity or difference is topologically and temporally restructured across some interface. It mediates different organizations of energy. The membranes of the microphone move in a magnetic field. A microphone couples soundwaves and electrical currents.

Simondon generalized the specialized engineering usage to give a transductive account of machines. He reinterpreted cybernetic theories of information and technology according to transduction, and at the same time sought to develop a richer notion of information. Unlike the cybernetic mainstream represented by Shannon, Weaver, Wiener and von Neumann, Simondon does not regard machines as *producers* or *consumers* of information, but as *transducers of information*. Following information theory, he understood 'information' to refer to the indeterminacy or contingency in a series of signals. (The basic idea here is that more information is communicated by an unpredictable sequence of signals than by a predictable sequence. Information theory is a way of quantifying the level of unpredictability of communications within a given context.) Devices transduce information, understood as a margin of unpredictability in a sequence of signals, into determined forms. Any device that retains a margin of indeterminacy can transduce information: 'information supplies determination to the machine' (Simondon, 1989a, 144). Information literally in-forms a machine, or imparts a form to it (or, at least, to that aspect of it which remains open to determination). When that happens, the device transduces different forms or organizations of energy. Thus in contrast to pervasive and deeply rooted modern models of mechanism which regard machines as deterministic, Simondon views the relation between a machine and its milieu as structured by localized and singular indeterminacies:

> The existence of a margin of indetermination in machines must be understood as the existence of a certain number of critical phases in their functioning. The machine which can receive information temporally local-izes its indetermination in sensible instants, rich in possibilities. (1989a, 144)

A machine, from this perspective, composes an ensemble of localized *suspensions* of determination, able at certain 'sensible' instants to receive information as a temporary and variable determination. It is not fully determined by any particular present since it maintains a margin of indetermination that allows it to cycle repeatedly through the critical phases. That margin opens technical action to a future. It projects into what will happen. No doubt a non-living technical object is still more or less located in the present, since it cannot transform itself outside those critical phases, 'rich in possibilities'. It must stand in relation to something other than itself in order to become something else. Yet, by virtue of its technicity, a technical object (especially a machine, but perhaps even more so an ensemble of machines) does not stand fully in the present. A device can transduce information repeatedly because it *suspends* its relation to the present or, put differently, because it retains a margin of indeterminacy through which it can keep receiving information without becoming an entirely different entity. By contrast, a less technical, perhaps non-living, object such as a rock tends to irreversibly absorb or undergo determination because it lacks the suspension or temporal localization of indeterminacy possessed by a machine.

4. The term 'life', especially in association with 'technology', is currently heavily overloaded. Making use of the term without reference to its complex

biopolitical dimensions is risky. Simondon's work does not acknowledge this situation to any great degree. However, his transductive approach remains significant. Life's transductivity is more complicated, and less linear, than the non-living physical individuation of a crystal. It is more differentiated than the transductions staged in technical ensembles since it entails growth, reproduction and usually death. Nevertheless, the notion of transduction destabilizes the hierarchy which assigns inorganic entities to a lower rung, beneath organisms. The living does not come *after* the non-living, but *during* it. From the standpoint of time, the living can be seen as a suspension of the processes of individuation that occur in non-living ensembles. Simondon writes that 'vital individuation would come to filter into physical individuation by suspending its course, by slowing it down, and by rendering it capable of propagation in an inceptive state' (1989b, 150). The contrast between living and non-living emerges through the delays, or desynchronizing processes that living ensembles unleash in themselves.

CHAPTER 1

Radical contingency and the materializations of technology

dependence of a chose

Can there be a theory of 'contingency' that is not compelled to refuse or cover over that which it seeks to explain?

Butler, 1993

Can technological processes themselves be seen as a source of contingency, rather than as something that covers over, limits or neutralizes it? Michel Foucault in his essay 'What is Enlightenment?' asked: 'in what is given to us as universal, necessary, obligatory, what place is occupied by whatever is singular, *contingent,* and the product of arbitrary constraints?' (1984, 45). This chapter draws on two streams of recent thought to steer that question towards engagement with technology: theories of corporeality developed by Judith Butler (1993), Elizabeth Grosz (1994) and others, and the extended critique of hylomorphic understandings of matter found in the work of Gilbert Simondon. These two streams refer the critical question formulated by Foucault to the contingencies of living bodies in their coupling with technologies. Some feminist theories of corporeality such as Butler's and Grosz's take Foucault's critical question and unravel its implications for universals such as gender and race associated with human subjects and collectives. These universals are given as natural, biological, instinctive, genetic or hereditary, and the lives of human subjects and collectives are seen as obliged to bear their imprint in different ways. According to these theories of corporeality, a reconsideration of the concept of matter could allow the contingencies of living matter to emerge and contest the necessity and universality of that obligation.

From a quite different angle, Simondon's fine-grained account of the processes of individuation complements body theory by showing how conceptually opposed terms such as form and matter, living and non-living, human and technical can be seen as abstract husks of the transductive interactions from which they derive.

If we want alternatives to a technological determinism which, in one form or another, and under one name or another (efficiency, profitability, national economic competitiveness, etc.) is 'given to us as universal, necessary', two problems arise. The first is to understand how technology is in fact *given* to us in this particularly hegemonic way. The second problem is to think how it could be understood as singular and contingent. On this second topic, I will consider what could be learnt from the theories of radical contingency discussed in corporeal theory. Within certain traditions of humanist thought, and certain strands of popular culture, modern technology, especially in its *informatic* dimensions, often figures (culturally, politically, economically) as a force of homogenization, dislocating and abstracting proper singularities. Cultural differences, in their ethno-geographic situation, are often seen as irreversibly changed by technological networks of communication and transport; genetic modification and biotechnology is seen to reconfigure universals associated with biological heredity. Computers, information and modern communications media have been frequently represented as neutralizing historical singularities in the name of a more universal norm: the functional efficacy of technology or, more specifically, *information*. While I will not be examining specific representations in any detail here, they are scattered throughout much science-fiction and film, they abounded over the 1990s in talk of information superhighways, and they continue to operate in public debates about technological change. Across all these different domains, the focus of this chapter is the ways in which technologization can be seen as contingent.

The apparent necessity of technologization begins to waver as soon as we articulate the situation of any technology in detail. Viewing technology as a homogenizing force is an extremely homogenizing representation of technology. Theories of corporeality can help to rework this homogenizing necessity at a theoretical level. Thinking through technology in association with the radical contingencies of embodiment might enable a different articulation of technological mediation. The powerful phantasmatic allure of disembodiment or re-embodiment associated with informatics and biotechnology in recent decades can be seen as one fetishized glimpse of the need for such a different articulation of technologies and living bodies. Crucial zones of indeterminacy, incommensurability and differences lie along the topologically and temporally folded boundaries between living bodies and the non-living organized matter we call 'technology'. What I am asking here is: how can we think about the radical contingency or finitude of embodied quasi-subjects *together* *with* their materialized quasi-objects, technologies?

TECHNOLOGY ABSORBS EVENTS

This problem of how modern technology is given to us has to be posed against a background meta-narrative. J.-F. Lyotard wrote that 'modernity is . . . a way of shaping a sequence of moments in such a way that it accepts a high rate of contingency' (1991, 68).[1] From this perspective, contemporary technological systems buffer and absorb higher rates of events. Ensembles of bodies, things, institutions, images and forces are subject to programming, in an attempt to render them calculable, predictable and tractable. Mass media, telecommunications, weaponry, genetically modified (GM) food and drug synthesis are the spin-offs of a process that accelerates events, as so many twentieth-century writers have described, by aggregating masses and groups of living and non-living bodies in programmed, repeatable sequences. According to the meta-narrative, this programming neutralizes differences and singularities, and allows them to be metabolized at a greater rate. Such programming even thrives on uncertainty and unpredictability as something to profit from.

Behind Lyotard's sophisticated formulation lies an important point. Important elements of Western thought, well represented in the history of philosophy, have long judged technical practices as derivative compared to thought, reason, subjectivity or substance. Well before technological modernity, technical practices were regarded as placing life, thought, ideas and subjectivity into risky proximity with the accidental contingencies of lifeless matter. In classical metaphysical terms, unless lifeless matter bears the governing imprint of living thought, randomness, chaos and exteriority afflicts the social-metaphysical order. This is clearly a severe oversimplification of diverse concepts of life, matter, body and thought lying behind contemporary technoscientific thought, but for the purposes of the modern meta-narrative, technology or 'the technical' cannot in principle be the source of anything radically new. Technological invention may reflect human ingenuity, but its novelty and singularity is secondary in relation to the properly human capacities such as living reason which conceive it, shape it and regulate its existence. The technical supports or supplements human life, but would never constitutionally affect who or what the human is. (It should go without saying that this position is untenable, and that a transductive understanding of the technical would begin by regarding this separation between human and the technical as the wrong starting point. Still, we need to account for the tenacity of this way of viewing technology.)

Contemporary technology continues to accentuate this theme of an accelerating absorption of events. The acceleration, as Lyotard's account, along with most twentieth-century critical theories of technology show, is often figured as catastrophic. While the catastrophe is usually figured as tending towards a greater and greater loss of human control (see Feenberg, 1991, for a broad account of critical theories of technology), Lyotard figured this loss hyperbolically in terms of an end to the conditions under which living humans bodies can survive, that is, the thermodynamic heat death of the sun. Lyotard's approach aside, there is also a widespread perception and even gleeful optimism that the end of the human as we know it is associated with technology. In this battle for primacy between humans and technology, nothing is really at stake because both sides accept that the logic of technological evolution is calculable and lawful. For both humanism and a techno-logically determinist posthumanism, nothing new eventuates through technology because its development is logical and predictable. In consequence, the response that has dominated much modern critical thought starts from the position that technology particularly threatens the passage of time by rendering it susceptible to calculation and prediction (see Chapter 4). As Lyotard, for instance, explains in his essay 'Time Today', the technological processing of contingency neutralizes indeterminacy in the present by programming what will happen in the future. It seems that for critical theory the inevitable logic of technological development, as well as the calculable, regular functioning of technology, forecloses the future. Only the superficial novelty of new gadgets remains. The present is constrained because the future has been at least partly shaped in advance through informatic modelling and organizational programming of actions and events. Technology threatens life and thought by wrenching it away from self-present experience and externalizing it in a lifeless or passive play of patterns.

In response to this threat, there is very often a nostalgic tendency to imagine that in another time and place, events would not be subject to capture, and singularities and differences would not be absorbed by technical mediations. This place has usually been located somewhere beyond technology, in institutions and living traditions which preserve values, meanings and symbolic interactions uncontaminated by the material processes of technical action. In European philosophy, contingency has been located in sources deep within the subject (Kant), in the historical dynamics of intersubjectivity and collectivity (Hegel), or in the remembering of being (Heidegger). In recent philosophy, the manifestations of this judgement have been divided. On the one hand, there has been an attempt to separate the dead, passive rationality associated with technology from a non-alienated, living reason (Marcuse, Habermas, *et al.*). Habermas' separation between instrumental-technical reason and social or communicative reason accepts this starting point (Habermas, 1987). On the other hand, Heidegger responds by developing a non-rational, non-objectifying or non-metaphysical thought of what exists, a way of thinking that does not seek to place technological reason under the control of a more proper reason, but to displace the primacy of reason altogether.

Despite deep differences, the position of technology itself remains the same: there is nothing radically contingent about technology. Whether seen as a juggernaut of progress, as the incarnation of dehumanized reason, or as the end of metaphysics, technology itself (whatever that might be) remains inert or passive. Heidegger's famous statement in his essay 'The Question Concerning Technology' that 'the essence of technology is nothing technical' (1977, 15) seems to repeat this same separation. In any case, most philosophy assumes that a hierarchical separation between technological and non-technological contingency is, in principle, possible. Technology, in its association with the senseless contingencies of matter, has been pushed outwards and away from the elevated interior spaces proper to life and meaning. In philosophy at large, reason, thought and indeed life are still seen to be more proper to humanity than technology.

LOCATION, DIFFERENCES AND TECHNOLOGY

Can we grasp technologies, in their very materiality and contingency, as something other than neutral instruments of human progress or implacable colonizers that exploit and master those characteristics that are most properly human?[2] Can we approach technology as locating and differentiating in its own way? Recent attempts by feminist philosophers to rethink what is at stake in sexually differentiated embodiment offer a starting point. Although technology does not always appear at the centre of theoretical discussions of embodiment and materiality, these theorists have a similar concern with the tension between contingency and necessity, singularity and universality, and between matter and form. (Feminist and cultural studies of science and technology have in recent years taken some of these theoretical insights into detailed historical and cultural studies of science and technology. Donna Haraway's work would perhaps represent the most powerful and sustained instance: see Haraway 1997; Balsamo, 1996; Plant, 1997. Haraway's work will be discussed in Chapter 6.) The problem which lies at the heart of these theories is precisely how to map contingent differences and locations in the constitution of subject-ivity without either essentializing those differences or reducing them to a pregiven form within culture. Theorists such as Butler, Grosz, Spivak and Gatens have done so, as Pheng Cheagh argues, by shifting 'from a model of independent subjectivity to an attempt to track the constitutive miredness of autonomous subjectivity in the always-already occurring momentum of a cross-hatching of hetero-determinations' (1996, 124). They seek to articulate the contingencies and location of subjectivity, without grounding them in universals of nature, reason or culture, by following their propagation across interfaces or sites of

differentiation where differences and indeterminacies elude solidifica-
tion and stabilization. The troublesome category of corporeality and
materiality allows them to ask: does the apparently meaningless
neutrality of matter already harbour unspoken or sedimented differentiations? Is matter, in its very appearance as malleable, plastic, fluid,
volatile or solid, already a sedimented reservoir of historical determinations? Such theories move, in short, from contingency understood as
something intransitive to contingency understood as bordering, touching and contaminating articulation of diverse realities with each other.
Corporeal materiality thus appears not as a substance, but as a pre-
eminently *transductive* field in which psychical, physical, technical and
affective realities precipitate.

RADICAL CONTINGENCY AND SIGNIFYING PROCESSES

Feminist work on corporeality has unfolded against a shared background of social, semiotic and political theory that has approached
questions of subject construction in terms of the structure of the
linguistic sign. Radical contingency figures largely in terms of productive breakdowns in signifying processes. Within poststructuralism,
radical contingency is not just an historical or empirical accident,
wherein, for instance, certain social formations undergoing seculariza-
tion according to technoeconomic imperatives lose all traditional
foundations or values. In the form of breakdowns or constitutive
failures in signifying processes, radical contingency inhabits the foundations of social formations. For many poststructuralist theories, contingency emerges from the very finitude of sociosymbolic or linguistic
processes, or at the limits of discourse, where signifying processes
break down. (Ernesto Laclau's (1995, 1996) theory of the empty
signifier would be one relevant specific example: the process of
signifying the limits of a social or political totality sets in motion chains
of signifiers which can never stand still because they respond to an
impossible injunction to signify the absence of any founding unity.)

An emphasis on breakdowns in signification is both suggestive and
limiting in relation to technical mediations. It usefully suggests that
contingency need not always be defined in terms of agency located in
either subjects or in a structure that itself acts as an agent. A system

of marks can generate dynamics, orderings, limits and centring effects in the absence of any pregiven ground or unity. When such a system encounters its own limits, transcendental or empty signifiers proliferate. The signifier 'technology', with all its associated affects, can be seen as one such limit term for contemporary Western collectives. It refers to no single signified or semiotic substance. It would be possible to map out how shifts in the signification of the term 'technology' over the last few centuries have allowed it to function as an empty signifier in relation to certain political and economic formations.

However, the emphasis on signification and its breakdowns as the sole locus of contingency usually stops short of questioning the materiality and temporality of technical acts, or it views them as always subordinate to the effects of signification or the structure of the sign. To bring contemporary technology into focus, to begin to articulate the tension between technology as homogenization and technology as singularity, the 'linguistic idealism' of much work in the humanities needs to be refocused in terms of matter and time.

MATERIALITY, EMBODIMENT AND TECHNICITY

Anti-essentialist feminist theory such as that of Judith Butler and Elizabeth Grosz has taken such a step by suggesting that corporeal differences trouble subjectivity with radical contingencies and singularities (Butler, 1993; Grosz, 1994). Bodies are more than an accidental contingency affecting subjectivity. They have sought to locate radical indeterminacy and contingency in embodiment, and specifically in the materiality of living bodies, rather than solely at the level of culture, language or socially constructed significations. Judith Butler's complex account in *Bodies That Matter* indicates how the metaphysical ordering which positions contingency, matter and bodies as the lesser terms of necessity, form, reason and language might be reconfigured.

Butler claims that *discursive practices* 'materialize' and 'dematerialize' sexually differentiated bodies. 'Matter', and specifically the matter of living bodies, is her name for a product of power, indeed, for power's most productive effect: matter for Butler is *not* simply surface and/or site on which social processes inscribe themselves, but the product of a materializing process which 'stabilizes over time to produce the effect of boundary, fixity and surface we call matter' (1993, 9). In a different formulation, Butler calls matter 'the sedimenting effect of a regulated iterability' (252). Butler shifts the stress from matter as inert ground to matter as an ongoing and variable effect, suspended in a web of interlacing processes whose general dynamics can only be understood in terms of iteration, citation and performativity.

How can matter be a product of power, stabilizing as corporeal surfaces and boundaries? In contrast to the more familiar notions of gender as a social construction, coding or inscription found throughout the human sciences, Butler is strongly attentive to the *limits* of discursive construction in general. Her approach moves away from either (a) assuming, along with humanism, pregiven subject positions which would employ themselves in making-over natural bodies into culturally significant entities; or (b) assuming impersonal structures such as 'discourse', 'power' or 'culture' which would act as agents in constructing human subjects.

Butler's notion of materialization as productive of matter takes a different and unavoidably more complicated path towards the limits of signifying processes. The solidity of structure, form and sign need to be augmented by reflection on their highly provisional grounding. Butler's account of materialization, standing at a singular confluence of the thought of Luce Irigaray, Michel Foucault and Jacques Derrida, shows how an apparently passive matter or natural materiality is itself fabricated in the heart of a material-social process of repetition and citation. While this process always involves language and power, it is not just that. That is, instead of either leaving matter outside the scope of social construction, or including it within the social as another socially constructed artefact, Butler argues that matter, as an ensemble of effects such as fixity, boundary and surface, is inextricably bound up with the determination of the limits and boundaries between the social or 'human', the natural or 'non-human', the constructed and 'un-constructed' and the living and the non-living. The insignificant or even senseless contingency associated with matter conceals the historically instituted sources of radical contingency that reside within the processes of materialization, within processes that tend to stabilize what counts as matter and what does not.[3]

In one of her most important arguments, Butler suggests that the effects of fixity or irreducibility associated with matter might stem from a discursive practice which cannot show itself as such without destabilizing itself and its products: 'If it can be shown that in its constitutive history this "irreducible" materiality is constructed through a problematic gendered matrix, then the discursive practice by which matter is rendered irreducible simultaneously ontologizes and fixes that gendered matrix in its place' (p. 29).

Her main interest here concerns how we should understand the endemic persistence of gender categories. Her response is to say that we can only understand their persistence if we take into account how the foundational term matter bears within itself a barely legible, encrypted version of those categories which serves to 'ontologize' gender categories that might otherwise display a great deal more heterogeneity. Instead of accepting the irreducibility of matter as the ground of human subjectivity, or strictly separating matter from the properly human, Butler delicately positions matter as the convoluted boundaries and pleated surfaces historically formed by processes of materialization – and dematerialization – governed by regulatory norms. Moreover, rather than simply arguing for the inclusion of whatever differences have been forcefully excluded by regulatory norms, the political task implied by this concept of materialization will be 'the preservation of the outside, the site where discourse meets its limits, where the opacity of what is not included in a given regime of truth acts as a disruptive site of linguistic impropriety and unrepresentability' (p. 53).

From this perspective, the boundaries and surfaces of what counts as matter are neither pregiven (as, for example, human biology), nor irrelevant (as, for example, the passive substrate of cultural constructions). They unfold in a domain or register where the distinctions between the social and the natural, the ideal and the material, are constituted and contestable; where regulatory norms and ideals cleave and divert different possible materializations, even as they rely on that very diversity of materializing repetitions to maintain their own regulatory force. By counting as matter, an encoded history of sexual and racial differences remains hidden. Yet the very fixity and receptivity of matter also asks a question: 'How is it that the presumption of a given version of matter in the effort to describe the materiality of bodies prefigures in advance what will and will not appear as an intelligible body?' (54).

Her answer to this question is complicated. Ideal norms, such as heterosexuality, which constitutes one important filter of intelligibility for bodies, function to dissimulate the fact that they can only persist as necessary or natural through their own *performance or repetition*. The power to materialize bodies according to sexed subject positions works through repetition and iteration. Norms are not origins or substances, but *institutions*. Furthermore, as Butler insists, 'repetition is not performed *by* a subject; this repetition is what enables a subject and constitutes the temporal condition for the subject' (p. 95). If repetition is primary, and the law is not an origin, then what appears as the irreducible matter of bodies, to be known, shaped and governed, is itself the reiteration of a contingent delimitation. The fixing, binding, bounding of matter acquires an appearance of passivity and receptivity through corporeal repetitions and citations that repeatedly and performatively exclude anything that does not comply with the norm in question. In other words, the matter of bodies is instituted, in all its ideal attributes – fixity, boundedness, etc. In recycling or refusing whatever does not conform to the ideal norms (for instance, of heterosexuality), matter solidifies.

The consequences of this performative understanding of matter are twofold. On the one hand, an apparently neutral limit or constraint on subject positions – the matter of a sexed body – is shown at least in general terms to enfold differences, and to be stabilized through historically instituted iteration rather than descended from an origin or essence. On the other hand, that which renders necessity contingent – iteration – is itself radically contingent because it is fundamentally improper, lacking in self-identity, mutable and incomparably generative. Iteration conforms to norms and ideal forms only because divergent events are constantly filtered out. Radical contingency, instead of being located only in breakdowns in signifying processes, folds back into matter through the positive and disruptive potentiality attributed to 'matter' as both limit point and 'the site of inscriptional space' in which all repetitions and performances eventuate (p. 52).[4] Through its improper citation or divergent performances of proper figures, radical contingency eventuates as matter. Stabilized through its 'inscription' of matter as site and surface (and especially the sexually differentiated morphology of human embodiment), radical contingency covers its sedimented history. Its show of stability requires the instituted system of norms and ideals which over time iteratively materialize the limits and boundaries of bodies; but at the same time it incessantly exceeds those limits, showing that the norms and ideals only appear to be necessary through their divergent citation or repetition.

ITERATION PROSTHESIS: NON-LIVING MATTER AND TECHNICAL PROCESSES

Butler's specific concern is to disentangle sexual difference from the 'sad necessities' (p. 53) of signification and gender, by rethinking materiality as radical contingency. Once she has shown how matter itself is always enfolded in a history of materializations, any norms and prohibitions that rely on the stability and neutrality of matter should take on a contingent status. They become liable to displacement or reorganization through the improprieties generated in citation. Her account of iterative performativity constitutes a corporeal answer to the critical question posed by Foucault. That question asks about the place of contingency and singularity in whatever counts as universal, necessary and obligatory. In Butler's analysis, we are given matter as 'universal, obligatory, necessary' only through the singular and contingent performances of living bodies in all their differences.

However, Butler's account remains focused almost solely on *living* bodies. What does such an approach to materiality mean in relation to technical processes? Does it open a way of thinking, even by analogy, about how a technical change correlates with differences and mutations in corporeality? If there has been a tendency to negate technology in philosophical thought (for example, Heidegger's 'the essence of technology is nothing technical'), it is partly because technology has been regarded as externally miming human (and non-human) action; it has been associated with the conversion of living interiority into the fixed, stable and uneventful forms of matter. As organized dead or non-human living matter, technology receives the imprint of human subjectivity and performs sequences of actions that flow from humans.

From this standpoint, if we mistake technical action for real action, we risk rendering human life passive, and losing the meaning of thought and action. The technological threatens life, both in thought and action, by performing too much or too well. Gadgets, machines and automata are a problem because they copy life. Recent debates about artificial life (a-life) for instance, oscillate between regarding a-life as a computer simulation, and a-life as a prototypical lifeform.

Following Butler's strategy, we could move away from this viewpoint. Like the iterative materializations of living bodies, technologies participate in materializing and dematerializing the limits, surfaces and borders of what counts as matter at a given time. Clearly, technologies participate in the instituted norms that Butler's theory highlights. However, technologies also introduce instabilities in the hierarchy of contingency, sometimes to the point of being mistaken for life (as in the long history of lifelike artifices and automata), but more often inconspicuously reorganizing living bodies by inflecting their bordering and synthesis of gesture and perception differently. The technological miming of life, in all its 'live' effects, its animations and artificiality, occurs through and as repetition. From a standpoint that sees it as mere extension or mime of the human, technology changes nothing; it formally repeats or supplements human action, and it organizes neutral matter.

Any mention of neutrality, however, should now alert us to the sedimentation of differences which it conceals. Butler's account enjoins us to say that neutrality is an effect produced by means which are never simply neutral. At a general level, corporeal materialization transpires through historically instituted normative iterations. By the same token, and in tension with corporeal iterations, the technological eventuates as an historical ordering and channelling of iterability in which living and non-living matter appears as capable of taking on form, of bearing an imprint or mark. If iteration shores up a clear line between form and matter, this iteration affects not only living bodies but organized matter, both living and non-living. If iteration has no origin, it is not solely (living) bodies that matter, but the body and its originary supplements. Living repetitions, in their divergence, are always touched by a non-living repetition.

TRANSDUCTIVE CONTACT

Such an augmented account of materialization through iteration might allow a different consideration of the contemporary technological acceleration of contingency than that which Lyotard offers. In general, he regards technological systems as tending to absorb contingency by laying down in advance the patterns within which events can be received. By contrast, at least in general terms, a focus on the materialization of technology suggests that the neutrality of the technical-material substrate already encodes singularities and differences. While the idea of technology as a neutral means or instrument of human action has long been subject to criticism (for a historical survey, see Mitcham, 1994, 230–1), the neutrality of its materiality in relation to human embodiment has remained largely unquestioned (with the possible exception of Karl Marx; see Markus, 1986). Butler's analysis points to the history of matter as that of exclusions and asymmetries. Technology participates in those materializations, and thus also in the exclusions and asymmetries they enact.

Butler's work suggests that matter as a category can be better understood as sedimented history than as irreducible, necessary, or ahistorical ground. The political motivation for such an account of bodies in terms of what counts as their very substance draws on the 'improper' citations of dominant social norms enacted in diverse ways by feminist, queer, and anti-racist movements. The deconstructive subtext of this theory concerns the instability and unlocalizable performativity of writing and marks. Although Butler's account focuses on differences and improprieties stemming from recognizably social contexts, the source of divergent repetitions is not limited to those contexts. Indeed, what counts as pertaining to bodies might be very difficult to limit. Gayatri Chakravorty Spivak writes, 'if one really thinks of the body as such, there is no possible outline of the body as such' (Spivak and Rooney, 1994, 177). Although this could be read in a number of ways, it primarily suggests that, ecologically speaking, there is no such thing as a body as such, by itself, in isolation. Read transductively, what we take to be a body, with its limits, knots together domains. This can be seen as a direct consequence of Butler's theory of the constitution of the boundaries and surfaces of different bodies. Unless we assume that a body has pregiven limits (for example, that a body is always fully alive), there is always a potential contamination of the living by the non-living, of the natural by the technical, of *physis* by *techne*. The inherently unstable and divergent advent of iteratively stabilized bodies cannot radically exclude the non-living. Distinctions between the corporeal and the non-corporeal, between the technical and the non-technical, and between the living and the non-living, cannot be exempt from this logic of interactive destabilization. Instabilities in the line between matter and form highlighted by Butler's theory can and do propagate through the associated network of conceptual oppositions.

If matter as such is historically sedimented, there is clearly no justification in limiting the focus to living matter. Non-living matter should also be involved in the process of establishing norms and laws.[5] Corporeal feminism's contribution to a material culture of technologies may well consist of asking how non-living materializations participate in the exclusions or asymmetries through which matter comes to stand as the neutral substrate of socially elaborated forms and imprints. Just as the political project of these theories aims to recognize and represent patterns of exclusion and violence concerning differences within the apparently neutral domain of corporeal materiality, the political stakes in a theory of technological materializations might be conceived as articulating the ways in which a persistent anthropomorphism negates its troublesome differences and normalizes its own production through the apparently contingent mutability of technologies. The very threshold between what does and does not count as technological relies on stabilized materialities, living and non-living. At least in outline, this approach sketches a response to the critical question of how technology is given to us as universal, necessary and obligatory. In practices and institutions (political, educational, economic and military), what counts as technology is part of an interactive stabilization of the human. The heavily asymmetrical investment in certain modes of technology (informatic, biotechnological, biomedical, communicational) can be read as symptoms of the way technology is currently given to us as universal. From this perspective, politics is in technology just as much as it is in the more visible and enunciative domains of collective symbolic interactions. The crucial political issues turn on how these domains are separated and enfolded with each other.

INFORMATION, FORM AND REPETITION

Following on from this still very general theory of technological-corporeal materialization, we can now address the second facet of Foucault's critical question: what place does the singular and contingent occupy in relation to the universal givenness of technology? The facet concerns the kind of analyses we might make of specific contemporary technologies. The obvious examples here are nearly always 'high-tech'. (That alone should give pause for thought. Why are so many other technical practices consistently rendered invisible or insignificant?) Over the last several decades, computer programs, information and networks have become the model for the technological organization of both living and non-living matter. The point has been made frequently that everything from biotechnology to contemporary mass media relies on the programmed citation of marks, on their analysis as combinatory sequences of discrete operations. This is a consequence of a complicated pattern of interactions which are still being mapped out (Haraway, 1997; Hayles, 1999, for a recent survey; Edwards, 1996, for a detailed North American perspective). The question that Butler's theory prompts in relation to the historical institution of information as the (currently) pervasive mode of organizing production, communication, form and meaning would be something like this: how does information stabilize the surfaces and boundaries of living and non-living bodies? What kinds of regulatory ideals for what counts as 'the human' contingently and historically emerge through informatic materializations? An account of the interactive contamination between non-living and living materializations and the specific iterability of information will be developed in the chapters that follow. It is not difficult to make the general deconstructively inspired claim that living bodies must always be already contaminated by their own prostheses, that they can only be what they are through their non-human extensions and supplements. It is a little more difficult to specify how the topological and temporal limits of bodies, living and non-living, are interactively stabilized and destabilized, especially in contemporary informatic milieus. So far, we only have a general account of the necessity of this interaction, of why it should accompany any questioning of the given-ness of corporeal matter. We still lack a way of approaching the constitutive coupling between the living and the non-living, between the technical and the non-technical, in information technologies for instance.

RECONFIGURING HYLOMORPHISM: A TRANSDUCTIVE UNDERSTANDING OF INFORMATION

Any such approach must circumvent or at least render problematic the form/matter or *hylomorphic* distinction. There are several reasons why hylomorphism should concern us. Many ways of thinking about subjectivity, culture and technology still adhere to that distinction in one guise or another. The distinction still often operates, implicitly or explicitly, as the paradigm of a technical operation. Not only matter, but many other substances are said to be formed in a technical act. Hence, social theory often conceives the interface between the human and the technical in hylomorphic terms: the human (collectively or individually) *shapes* or is *shaped* by technology. Technology is seen as a way of forming energy or matter in the interests of human life. This secondary status is often mirrored in contemporary debates concerning the digital embodiment of information. Here digital information is seen as a form independent of its material substrate (whether that be optical fibre, a silicon chip or a laser-inscribed plastic disc). Similarly, genetic information coded in DNA sequences has been treated as independent of the complex rhythms and topologies of organisms (Hayles, 1999, Chapter 1).

The work of Gilbert Simondon suggests a different notion of information. It stresses the diverse relations mediating the transparent and inert terms of form and matter.[6] This alternative account of information allows us to develop the notion of transduction discussed in the previous chapter and, at the same time, formulate a response to the question of the place of singularity and contingency in relation to the global universal of technology. The account unfolds on a number of levels. Simondon resists hylomorphic understandings because the notions of form and matter remain too abstract and static to grasp the convergent chains of transformations unfolding during the technical action of moulding in which matter takes form. Making a brick, for instance, might seem trivial, and profoundly different to information processing, but Simondon's understanding of brick-making is typical of the way he reconceptualizes the opposition of form and matter as a transductive process, or as a process of *information*. According to Simondon, the existence of a brick presupposes a carefully mediated encounter: 'In order that there can be an actually existing individuated parallelpiped [i.e. a solid bounded by parallelograms] brick, an effective technical *operation* must institute a mediation between a determinate mass of clay and this notion of the parallelpiped' (1989a, 38). Moulding the clay into a brick depends on a prior set of operations, involving preparation of the mould, and preparation of the clay. The brick, as a technical element, is a domain in which different realities have been transduced or mediated:

> The mediation is prepared by two chains of preliminary operations which cause matter and form to converge towards a common operation. To give a form to clay is not to impose the parallelpiped form on the raw clay. It is to tamp down the prepared clay in a fabricated mold. If one starts from the two ends of the technological chain, the parallelpiped and the clay in the quarry, one has the impression of realizing in the technical operation an encounter between two realities of heterogeneous domains, and of instituting a mediation through communication between an inter-elementary macrophysical order, larger than the individuated, and an intra-elemental, microphysical order smaller than the individuated. (p. 38)

Making a brick is transductive. It links 'realities of heterogeneous domains'. The technicity of brick – its durability, resistance to weathering, capacity to bear certain kinds of load, the bond that mortar can make to it – emerges from the mediation of different domains. The capacity of the material to be moulded is itself the outcome of a series of transformative operations. The clay must be prepared, for instance, so that it is homogeneous (large pebbles will disrupt deformation in the mould; Simondon terms them 'parasitic singularities' (p. 42)), plastic (that is, it will not shatter like glass when it is pushed down into the mould, but will simply deform into a different shape), and yet able to maintain consistency so that it can take on contours without spilling like water. All of these properties can be understood as 'intra-elemental', or 'microphysical'. The preparation of a mould is also technically complex, since a geometric form must be stabilized. It needs to receive and limit the deformation of the clay without making it difficult to remove, it must be strong enough to withstand the pressures of tamping, yet flexible enough to allow brick to be taken out without sticking. The mould presents an ensemble of moulding gestures, frozen or immobilized.

What is usually considered as single act of the forming of matter is only the last episode in a series of convergent transformations. Even in that last episode, the mould, as Simondon points out, does not impose a form from the outside on an inert matter: 'the mold limits and stabilizes rather than imposing a form: it gives an end to the deformation' (p. 40). In the 'common operation' where form and matter encounter each other under pressure, the mould solicits a state of internal resonance in the mass of clay. During the brief interval of moulding, the molecular elements of the clay redistribute the potential energy they receive from the gestures of tamping by filling the mould, pushing against its walls and then realigning themselves. The walls of the mould limit the microphysical ordering of the clay, but only through a point-by-point application of opposing forces to the expanding surfaces of the mass of clay.

Leaving aside the complications of variable moulds (although these complications are precisely what we will need to consider in relation to electronic information; as, for instance, in the encounter between gestures, word-processing software and display of text occurring as I write), the mould supplies conditions under which a reciprocity of interactive forces within the clay occurs. For only an instant, the moment at which the informing occurs, both mould and the whole mass are in dynamic relation. In the system composed of limiting mould and homogenized clay under pressure, forces propagate reciprocally throughout the mass of the clay, not just across the interfaces between clay and mould. This momentary metastability is transductive. It is an individuation in process. Simondon writes:

[M]atter in the process of taking form is in a state of complete *internal resonance*; what takes place in one point resonates with all the others, the becoming of each molecule resonates with all others, at all points and in all directions. . . . The mold, as limit, is what provokes this state of internal resonance, but the mold is not that through which the internal resonance is realized. (p. 43)

The interaction between mould and clay during the instant the clay is packed in and tamped down settles into equilibrium when the pressure exerted by the fixed surfaces of the mould equals the intra-elemental pressures propagating back and forth in the clay. The materialized form and the prepared material interact through a set of energetic exchanges which transform the potential energy of the clay under pressure (due to tamping) into a stable, determinate equilibrium. The forming of the matter does not only involve surfaces; it involves the whole mass of prepared material in a topological and energetic redistribution of the potential energy contained in the compressed clay. Only then can a brick be turned out of the mould.

INFORMATION AS ITERATIVE TAKING-FORM

The basic problem with the hylomorphic scheme is that it only retains the two extreme starting points – a geometrical ideal and formless raw material – of a convergent series of transformations, and ignores the complicated mediations and interactions which culminate in matter-taking-form. Without taking account of those transformations and their encounter, there is no way of understanding how the modulation of material by a limit-form occurs. The intermediate articulation of two divergent realities, the macrophysical materialized order of the mould, with all the gestures it immobilizes and perpetuates, and the prepared microphysical order of the clay with its homogenized distribution of colloidal particles, whose relations to each other allow deformation, remains invisible.

However, even if we recognize that the form-matter distinction conceals a complicated series of energetic interactions, it has another limitation. Within the domain of technical practice, moulding is a highly specific operation, performed by ceramic artisans for instance, but not easily generalizable to other technical operations. Metallurgical work, while it does mould its products, submits alloys to other transformations (quenching, annealing, etc.) which cannot be readily understood as matter-taking-form. Weaving or spinning also cannot be understood in these terms. We could in fact, according to Simondon, regard moulding as a limit case because the process of transduction occurs only during the instant in which matter and form constitute a single metastable system of interactions (p. 55). After equilibrium is reached, the brick is 'individuated', and the interaction is complete. The mould and brick diverge again as the brick is released from the mould.

This leads to the second level of Simondon's alternative to hylomorphism. Hylomorphism provides no way of accounting for or even acknowledging ongoing processes of formation. The momentary transduction is not repeated or sustained when matter-takes-form. If, for instance, we wanted to see how the iterative stabilization of corporeal surfaces and boundaries in Butler's work could be supplemented by this more transductive understanding of matter-taking-form, we would still want to ask how the performative materialization of living bodies continues over time. After acknowledging that matter-taking-form is much more complicated than form being given to an inert passive material, the problem of understanding the *iterative* sedimentation of living bodies, in their corporeal-technical specificities, remains. In what ways is ongoing materialization driven?

LIVING INFORMATION

In order to cope with this difficulty, the hylomorphic schema must be supplanted by a notion of *information*. Hylomorphism presumes that matter is given form only once or at one time. It ignores, moreover, the variety of interactions involved when matter is in formation. By contrast, information, at least as Simondon understands it, occurs whenever a transductive event establishes an intermediate level at which disparate realities can be articulated together. Information in this sense does not arrive from outside a system and need not be a discrete event. It eventuates whenever the 'unresolved incompatibility of a system becomes an organizing dimension in the resolution' of that incompatibility (1989a, 29). This unresolved state can endure. Information in this sense (and we will need to compare this concept with technological or informatic understandings of information) implies ongoing transductive processes in the living and the non-living. Without moving away from the central insight that brick-making provides, the suspended resolution of matter-taking-form allows for an ongoing process of individuation.

The relation between bricks and living bodies becomes clearer. Since living entities individuate continuously, rather than being formed once, they *are* information. They are continuous, variable processes of matter-taking-form. The instant in which clay and wooden mould exchange energy is prolonged in life. Life continually resolves incompatibilities through recurrent processes of 'internal resonance' resulting from the encounter of divergent orders or different scales. So for instance, a perception and a gesture participate in a process of information when they resolve a problem for an organism in search of food. The association between information and 'internal resonance' stresses that individuation does not occur because an entity is in relation to something else (to an exterior milieu, for instance), but because it is the 'theatre or agent' of an interactive communication between different orders (p. 60). In other words, information is a transductive process which provisionally resolves some incompatibilities within an ensemble. It may do so irreversibly, as in the case of the brick, but it may also do so by suspending or delaying formation, by maintaining or continuing the processes of formation, so that the ensemble remains information, as in the case of life.

It might be thought that Simondon's concept of information, and its appearance as a critical alternative to hylomorphism, bears little relevance to contemporary technologies of information and communication. Information as matter-taking-form, however, directly confronts the technical concept of information developed by cybernetics and the mathematical theory of information during the middle of the twentieth century.[7] Those theories can reinforce certain hylomorphic assumptions. If information is understood simply as coding, then what is coded appears as matter to be formed. Information processing technologies appear to form or shape what they process. Yet cybernetic information can be viewed transductively. Information in the cybernetic or informatic sense 'is not form, nor an ensemble of forms, it is the variability of forms, . . . the unpredictability of a variation of forms' (Simondon, 1989a, 137). Strictly speaking, information in the cybernetic sense measures the degree of uncertainty associated with the arrival of any signal. In flipping a coin, the possible signals are heads and tails. The information associated with a random coin toss is, to use the accepted conventions, expressed as 1 bit (mathematically, I $= \log_2 n$ where I is the quantity of information, n is the number of different possible signals, in this case 2 — heads or tails. Katherine N. Hayles provides a very accessible account of cybernetic information: see Chapter 3 of Hayles, 1999). Once the coding is established (for instance, that a coin toss has only two possible outcomes, 'heads' or 'tails'), the number of different states in a signal system can be calculated. This way of measuring information does permit calculations comparing different ways of encoding signals, but it does not pro-gramme the arrival of any particular signal. It is not so much the much-maligned divorce between signal and meaning that Simondon resists, but the confinement of the concept of information to a relatively late phase of its emergence, the advent of a particular form. (The implications of this very general point will be developed in later chapters. See for instance, Chapter 5, which discusses a particular informatic ensemble, performance artist Stelarc's *Ping Body*, Chapter 6, which discusses a real-time online computer game and Chapter 7, where genomic databases are discussed.)

BETWEEN CORPOREALITY AND TECHNICITY: TOPOLOGY AND TEMPORALITY

Judith Butler's corporeal theory has dramatized the radically contingent materiality of living bodies. Gilbert Simondon's alternative account of information augments this scene with a finer-grained treatment of the temporal and topological complexes in which differences between living and non-living bodies precipitate. Information in Simondon's sense has both a general and a particular scope. This way of understand-ing how matter takes form, while it remains general, opens a different path into technical-corporeal materializations and the complicated sites of interface where materializations occur. At a specific level, it

responds to the prevalence of information and communication technologies which have radically reorganized human collectives over the last
50 years. From this angle, we can treat this re-reading of information
as attempting to slow down the process in which matter takes form
through information technologies.

In corporeal theory to date, investigation of the norms governing
what it is to be a recognizably human subject have concentrated on
introducing living bodies into accounts of subjectivity and culture. The
interplay with *non-human* bodies and *non-living* ensembles could be
better understood. The explanatory weakness of hylomorphism stems
from its reductive treatment of matter as limited by form. Through a
transductive understanding of information, we can begin to see how
limits and boundaries between matter and form are interactively
stabilized. From the perspective of this interaction, technologies are
not a domain exterior to human bodies, but are constitutively involved
in the 'bodying-forth' of limits and differences. Technical materializations are always involved in what we take to be a living, human body.
Simondon's account highlights the fact that materialization is topologically and temporally complex. Transductive processes occur at the
interface between technical and non-technical, human and non-human,
living and non-living. We saw that during the process of information,
as two chains of operations encounter each other, mould and material
cohere in an internally resonating ensemble. If this event is somehow
prolonged, or its resolution is suspended, then the process of information becomes even more topologically and temporally complex. Contours, surfaces and limits may well be formed, but only provisionally,
and only as part of an ongoing process of individuation. If we seek to
comprehend what happens between humans and their non-living
prosthetic supplements without presuming we know the essence of
either prior to their encounter, and if we seriously regard human
collectives as co-individuating with technologies, then the topological
and temporal complexities of information should be taken into account.

At a more particular level, Simondon's notion of information acts
as a countermeasure to the tendency of recent cybernetic and biotechnological understandings of information to collapse living and non-
living processes together. It takes the specificity of machines and life
seriously. Machines are in the process of in-formation. They are open

to information to the extent that they can maintain a margin of indeterminacy, or a capacity to be in-formed. Within localized limits, the mechanism of a machine *suspends* final determination of its own form. Machines sometimes possess a rich repertoire of forms. A fully determined mechanism would no longer be technological; it would be an inert object, or junk. A machine must articulate some degree of openness to a milieu in order to remain technological. Conversely, a machine, no matter how sophisticated in its computational architecture, is not open to just any event. It is certainly not fully exposed to random events (although such a machine would be of great interest!). A machine works within a certain margin of indeterminacy maintained at its interfaces. This margin, which permits it to repeatedly be informed, and to be linked together in ensembles, is a precondition of 'information processing' as it is usually understood in contemporary technological processes. In preserving a margin of indeterminacy, technical artefacts, machines or ensembles allow themselves to act transductively. That is, they furnish a scene in which repeated energetic interactions between living and non-living bodies can occur. (Life too remains in-formation, but with the added orders of complexity that flow from its collective entanglements and from the internal milieus it maintains.)

Without wanting to pre-empt the complexities of contemporary information systems, word-processing provides a simple example of this indeterminacy. Instants of determination of the state of the machine (as keys are pressed on a keyboard, for instance) occur through the intersection between gestures and the temporal and topological organization of electrical currents. Selection of a form (a keystroke propagates through a hierarchy of coded sub-ensembles until very soon it appears as an arrangement of pixels on screen as a font character of specific size and style) occurs when two series of operations converge, those of a user habituated to keyboard and screen, and the layers of inscription mobilized by the system. The openness to determination of a machine brings all the complexities which the account of matter-taking-form has alerted us to. On the side of the machine, each keystroke entails matter-taking-form. Synchronized several hundred million times per second, an energetic configuration of potentials takes shape or actualizes itself as marks visible on the screen, held in memory or propagated into other technical ensembles.

The question this chapter opened with was whether technology could itself be seen as the source of contingencies in its own right, or whether in accord with the standpoint that has dominated most humanist responses to modern technology, it could only be seen as absorbing contingency at an ever increasing rate. From this latter standpoint, the accelerating effects of speed and change only signal that nothing new is happening. How would that opening question be answered now? Moving back a little from the details of Simondon's account of machines and in-formation, it becomes clear that the notion that machines maintain a margin of indeterminacy has important implications in relation to the above question. There is a margin of contingency associated with technical practices or mediations. This margin is transductive. It articulates different orders of reality. Rather than any particular technology determining the form of human life, it could be seen as suspending the final determination of form within localized margins. This is not to claim that technology is a locus of radical contingency or pure novelty. In this chapter, I have to tried to suggest that a different approach is possible. The deconstructive motif present in Butler's theory of iterative materiality indicates that what we take to be the most inert, passive term, matter, stabilizes only through processes in which divergent embodiments are repeatedly aligned or normalized. The sources of radical contingency do not reside in the subject or in the predicates of consciousness, nor in any body as such. Rather, they stem from the limits of thinkability of bodies, from the ways in which they have no possible outline or form. The a-signifying status of matter, indeed its very existence as an isolated term, can then be understood as a residue of historically interwoven institutions, practices and discourses through which inherent corporeal divergences are realigned and held in tension. Subjects – white, black, woman, man, homosexual, heterosexual – result.

The materiality of technology itself should be examined together with that of living bodies. It too must be seen as historically sedimented, and as instituted iteratively and through differences. The place of 'whatever is singular, contingent and the product of arbitrary constraints' (the focus of Foucault's critical question) is shared across living and non-living entities. Various implications flow from this understanding (and will be developed in the following chapters). The iterative normalization of bodies need not be narrowly construed as ascribing meanings or values to bodies. Value codings gain their significance only in conjunction with technical performances which provisionally resolve compatibilities in and between living and non-living processes. Simondon's response to hylomorphism offers one way of beginning to delineate those performances, and to glimpse something other than a relentless obligation to take up technology or to become more technological. The stabilization over time of bodily boundaries and surfaces need not be seen as either imposed from the outside (as a form), or as essential to bodies, but as the consequence of a 'common operation' occurring between the living and the non-living. The connection between the two approaches is not yet clear. My argument will be that if we can understand corporealization or materialization transductively, then it might be possible to ask how corporeality is already technical.

NOTES

1. This view of technology as an ensemble of arrangements for the processing of contingency diverges from the more usual views of modern technology as either instruments for the rational domination of nature and others, or the view of technology as some kind of extension of the human organs, as something of human making.
2. In general terms, Paul Virilio's work strongly represents this position. See Chapter 3 for further comments on Virilio.
3. For instance, one must be either a man or a woman, or risk exclusion from sociality (through psychosis or failure to materialize as a subject). This imperative is predicated on materiality, since matter, and particularly the sexual matter of human biology, is regarded as the necessary, irreducible, prediscursive ground of sexed subjectivity.
4. The place or support of inscription can itself never be inscribed or figured as such because it is the condition under which forms and figurations materialize. Beyond the opposition between active form and passive matter, it is the domain within which that ideal opposition maintains itself. It is a site of 'linguistic impropriety' in which the apparent necessity of the delimitations and determinations of matter can be re-mapped as contingent violence (Butler, 1993, 53).

5. Recent social studies of science and technology have made serious efforts to redress the asymmetries in analysis of human and non-human actors. Actor–network theory, associated with Latour (1994a, 1999) and Law and Callon (1995), is predicated on balancing the human and the non-human. This point will be more extensively discussed in Chapter 2.

6. Later chapters engage with the debate in various forms. See Chapter 3 in relation to the embodiment of digital information, and see Chapter 6 in relation to bioinformatics and its treatment of genomic information.

7. Simondon's response to cybernetics appears in a number of different works, but especially in Simondon, 1989a, Part II, Chapter 2.

From stone to radiation:
the depth and speed of technical embodiments

> For the thing we are looking for is not a human thing, nor
> is it an inhuman thing. It offers, rather, a continuous
> passage, a commerce, an interchange, between what
> humans inscribe in it and what it prescribes to humans. It
> translates the one into the other. This thing is the nonhu-
> man version of people, it is the human version of things,
> twice displaced. What should it be called? Neither object
> nor subject. An instituted object, quasi-object, quasi-
> subject, a thing that possesses body and soul indissociably.
>
> Latour, 1996

If we wanted to 'experience' contemporary technology differently, the
task would be not only to become aware of how discourses and affects
(for instance of urgency and necessity) cluster around technology, but
to find ways of affirming the historical or genealogical contingency of
technical mediations. The general lesson drawn from corporeal theory
in Chapter 1 concerns how the limits of discourse are entangled with
what we take to be bodies. Just as corporeal theory has made it
possible to ask whether living bodies, which we often take to be
outside or at the very limits of culture, inhabit a necessarily somewhat
occluded historical dimension, it might also be possible to ask how
technical mediations participate in the historical institution of collectives. That is the major focus of this chapter. However, there may be a
limit to our ability to represent technical mediations. It may be that
we cannot experience technology in general or a specific technical
mediation as such, any more than we can experience the body or our

own bodies as such. As we have already seen, the transductive event of a technical mediation works along other lines than the established semiotic and value codings available within a social system. It has, in short, a different mode of historical existence to a sign, even when it remains, as is often the case, closely coupled to signifying practices.

The currently dominant inscriptive practices of informatics close the distance between signifying processes and technical mediations. Because technical mediations such as computation and biotechnology are so entwined with signs, it might seem that they could be represented fully. As Bruno Latour writes, 'programs are written, chips are engraved like etchings or photographed like plans. . . . But then is there no longer any difference between humans and nonhumans? No, but there is no difference between the spirit of machines and their matter, either' (Latour, 1996, 223). Discursive limits are entangled with technical mediations, just as much as with corporeal performativity. Drawing on theoretical insights developed in recent social studies of science and technology, this chapter elucidates something of the historical depth of technical mediations. At the same time, it draws attention to certain constitutive limits in our capacity to represent and experience technology. In order to weave the general point drawn from corporeal theory together with these insights, it relies on a quasi-technical allegory concerning two technological stereotypes: stone axes and thermonuclear bombs. Highly dubious in terms of any strict philosophical, historical or social scientific investigation of technology, the allegory serves to highlight the problem of the depth of technical embodiments along transductive lines.

The story told here resembles the opening to Stanley Kubrick's film, *2001: A Space Odyssey*. There, a bone thrown by a hominid ancestor arcs up into the air in slow motion, and becomes a space shuttle carrying Dr Floyd towards an encounter with information technology, in the guise of HAL9000, a dominating supercomputer. The transition between the proto-technical act of throwing and the complexities of that now-past future occurs almost seamlessly in the film. What happens during the moment of inflection when bone becomes rocket? The slight break in continuity asks us to associate two technical artefacts as different from each other as possible (one signifying slowness and simplicity, the other, intimidating complexity

and power) with the presence of a black monolith, an incomprehens-
ibly alien technology that remains stable throughout millennia.[1] My
allegory is only slightly less mythical: it tells how an artefact from the
earliest human–non-human collectives, the Acheulean hand-axes of 1.5
million years ago used by *homo erectus*, one of the earliest technical
artefacts associated with our hominid ancestors, becomes something
that stands at the limits of contemporary technical mediations in terms
of its disproportionately lethal efficacy – the thermonuclear weapons
designed, stockpiled, detonated and above all, *simulated* by nation-
states after World War II. This is a story of artefacts at the limits of
discourse or signification.

These axes, the Acheulean, are located at one limit of the human.
They point to a time of mythical origins when we were becoming who
we are socially, linguistically and corporeally. They accompany an
evolutionary phase of humanity associated with massive corporeal and
presumably collective reorganization. (For a description, now somewhat dated but still fascinating, of this process, see Leroi-Gourhan,
1993.) While these stones, found scattered across at least three
continents, will not speak with complete intelligibility, they might be
readable in terms of originary technicity; that is, as something entangled with any idea or representation of what it is to be human. At
another extreme, the thermonuclear weapon constitutes a kind of
discursive limit for certain collectives today. Its detonation remains an
exceptional event that can only be seen at a distance on the horizon,
as a blinding flash of light and, even then, inevitably as a highly
mediated image drawn from stock film footage taken at a time when
atmospheric tests were still being conducted. Most importantly,
because it still might have to be detonated even after the end of the
Cold War, the bomb has become the object of intensive simulation.
Large research institutions and the most powerful supercomputers on
the planet are devoted to it.

TECHNICAL MEDIATIONS AND DISCURSIVE LIMITS

The choice of these two artefacts is not innocent. As a technology
subject to massive simulation, the nuclear bomb, perhaps more than
any other technology except computers themselves, is also a weighty

semiotic object.[2] As Paul Edwards writes, 'the Cold War can be best understood in terms of *discourses* that connect technology, strategy and culture: it was quite literally fought inside a quintessentially semiotic space, existing in models, language, iconography and metaphor, embodied in technologies that lent to these semiotic dimensions their heavy inertial mass' (Edwards, 1996, 120). Nuclear weapons were and are surrounded by massive simulations because no one quite knows what would happen in a conflict fought with them. Nuclear weapons stand as a kind of discursive limit for contemporary technologies, and their polarizing influence on the Cold War still propagates many second order effects in the domain of cultures, technologies and politics. In their own way, hand-axes share many of these properties. They too are objects of simulation, since we cannot know what their effects meant. They too can confirm certain notions as to the limits of the human, and the distinctions between human and animal life, but also unsettle them.

Consider the following parameters of the two artefacts:

Durability: the hand-axes were made 1.5 millions years ago and apparently were 'used' for a period of 1 million years (Ingold and Gibson, 1993, 337). They are found relatively intact today in large numbers and in stereotypical form across three continents. What they did, how they were used, the difference they made, their involvement in human collectives, has left few traces apart perhaps from some marks on bones. Also found buried across at least three continents in bunkers and missile silos, and roving around the oceans, thermonuclear devices have been made since the 1950s. In themselves, they are not durable. Due to the half-lives of the isotopes they contain, which decompose at around 5% per annum, they have to be continually replenished with fresh reactive material. Although they are not stable by comparison, their effects, should they detonate, would endure over several hundred thousand years. In that sense, as Michel Serres points out, their performance constitutes 'the most lasting' of technical mediations or quasi-objects available to us (Serres, 1995, 90).

Complexity: a gesture (throw, blow, etc.) propels and guides the hand-axe. Even leaving aside the complications of missiles, guidance systems, nuclear weapons laboratories, communication and control networks that surround the device, and the jittery politics that launch them, approximately 4000 components in the thermonuclear bomb are involved in a controlled implosion that triggers the rapidly fluxing neutrons of the fission and fusion reactions whose energy release amounts to an explosion.[3] The technical performance of the two artefacts diverges correspondingly between that of the hand-axe whose technical performance can be enacted by humans at many different times and places (as evidenced by their distribution), and the bomb whose performance is only effective at different times and places arranged and supported by a vastly more elaborate ensemble, supplied by collectives on the scale of nation-states. There are far fewer bombs than hand-axes. Furthermore, the detonation of bombs is now over-shadowed by the simulation of their detonation.

Speed: the flight time of the axe is several hundred milliseconds for a path of 5–10 metres, and less if the axe is not thrown, but held as a hand tool. Detonation time lasts a millionth of a second for the bomb. After the end of the chain reaction in the deuterium, the propagation of electromagnetic, atmospheric and heat shockwaves will take slightly longer, but the speed of the chain reaction itself exceeds that of the fusion reactions occurring in the sun.

Flowing from considerations of speed, the *window of control* is the most important contrast for my allegory of discursive limits. A throw depends on timing. For the hand-axe, the window of control ranges between one and several tens of milliseconds depending on the length of its trajectory. That is, assuming that the hand-axe is thrown, it must be released at the right moment ± 10 ms (milliseconds) from the hand during the throw if the device is to hit a small target five metres away (Calvin, 1993). The problem here is that the neurones twitch. They can't modulate movements with any great accuracy. The timing jitter for spine-motor neurones is approximately 11 ms (Calvin, 1993, 246). On average they vary that much in their activation time. Neural feedback from arm to spinal cord and back at its fastest still requires approximately 110 ms. A problem of control develops because the window of control is less than the average variation in activation time, let alone the time of neural feedback. Technical performance, if it is to have efficacy, must be much faster than certain raw facts of our own physiology seem to permit. A kind of *race condition* is involved in coping with the brevity of these intervals.[4] The contrast with the window of control for the nuclear bomb is massive. It only lasts around one hundred millionth of a second. Unless the unfolding series of events within the casing of the bomb is timed to within a hundred millionth of a second, the devices behave unpredictably in the flash of detonation or, better, their detonation fails. To fabricate thermonuclear weapons that provide predictable, large and efficient yields of energy (and nation-states want that, as symbolic performances of their own executive potency), the interaction of those 4000 components or so in the first millionth of a second must be tightly controlled. Again, the problem is how to control movement, given that the crucial interval of mediation is so brief.

The thermonuclear reaction presents an event of exceptional brevity or high speed compared to the movement of the hand-axe. The event of detonation occurs within the topology of those 4000 components situated in relation to each other. The timing of the hand's release of the axe stems from hand–eye coordination, whereas the timing of the movement of components in the bomb reflects an expanded and more complicated field whose coordination requires something like the resources of a large nation-state.

PROBLEM: TOO MUCH SPEED?

The allegory should already make one thing clear: the problem of speed is not new. But how are these contrasts in timing and complexity and this problem of handling those excessive speeds to be understood? How should the story continue? Should the differences between these two technical mediations be explained in terms of the differences between premodern and modern technology, which would very soon take us into accounts of modern subjectivity, reason, technoscientific rationality and perhaps even alienation and the like?

For the moment, let the story continue along these lines (and it should be clear already that this too serves only to introduce a different ending, one that I prefer). The contrasting numbers and times gesture towards the problem of the speed of technical action. The problem, stated simply, is this: as a technical performance, the nuclear bomb has too much speed. The detonation travels too far and fast, it is too levelling of differences, it destroys and dehumanizes. Only madness would countenance using it, let alone building it. The bomb therefore marks excessive technological speed.

The most readily available response to the problem of excessive speed available in the humanities would be to inscribe a line or break between the hand-axe and the nuclear device, to break down technical mediations into two categories, the modern and the premodern. To take only one example among many, a starkly drawn version of the break runs through the work of Paul Virilio: 'With acceleration there is no more here and there, only the mental confusion of near and far, present and future, real and unreal – a mix of history, stories, and the hallucinatory utopia of communication technologies' (Virilio, 1995a, 35). Here the contrast between 'now' and 'then' rests on acceleration, or changes in speed. In terms of this break as it is usually understood, the way to understand the problem of excessive speed would be to take the two artefacts – the hand-axe and the thermonuclear device – and say one is premodern and the other is modern.

On this account the difference in speed between them marks a radical break and difference in kind. One is closely tied to the hand, it remains proximate to the body, to a small repertoire of technical gestures, and to a limited set of relations and signifying processes within a social collective. Its unchanging form over one million years refers to a stability that might be understood as involving both low velocities and minimal experience of speed. So it is *local* and *slow*. By contrast, the other is *global* and *fast*, insofar as it is modern. The staging of a detonation over a few nanoseconds draws on scientific theories, calculations and sophisticated apparatus. The great velocity of the detonation, and hence its potency as a symbol of sovereign superpowers, relies on this technoscientific understanding and manipulation. It implies something much more extended, and much more global in its scope than the hand-axe. The high velocity of the detonation, the millionths of seconds, are carefully modelled and planned through calculation. The unrivalled speed of the chain reaction between the hydrogen nuclei is calibrated through the calculations concerning nuclear and thermodynamic processes carried out in the nuclear weapons laboratories in Berkeley, Los Alamos, Saclay or Aldermaston.

The great gulf that we customarily accept between the hand-axe and the bomb, the gap between the premodern and modern, assumes that one of them is valid globally, that one of them is not the work of local culture, but the work of a culture with the capacity to project beyond itself, to cut its ties to local contexts and customs. We habitually attribute speed to subjects whose access to the dividends of scientific *rationality* allows them to take a universally valid 'world view'. Any notion of radical break between the premodern and the modern holds as its axiom that, for better or worse, the modern viewpoint is more encompassing, more substantial and translatable than the premodern, that its claims to universality outdistance and overtake the particular forces of local cultures. The split between the modern and premodern then allows a role for the local only in the gaps left by the calculations, only in the zones that have not yet been translated into the more generalized system of technoscientifically planned action.

According to this viewpoint, anything that is thoroughly modern, such as the thermonuclear device, can only be modern to the extent that it is universal, or insofar as it embodies something that spans all localities. Machines go to Jupiter and Mars – sometimes – on the strength of that universality. Nuclear weapons capitalize on the fundamental energetic relations of sub-atomic particles. Although its *symbolic functions* for collectives are ancient (brandishing symbols of sovereignty is an old practice), the technical efficacy of the thermonuclear bomb, its operational capacity measured in explosive megatons of TNT, is thoroughly modern because it derives from universal laws, captured and channelled through technoscientific representations back to nature. From this standpoint, this technical action is bound to delocalize itself, even if its sociosymbolic function remains linked to particular nation-states, because it rests on the prior dislocation entailed in formulating a physical theory of the universe. The local, including embodied subjects, must then be figured as a remainder, as a residue of whatever has not yet been captured and translated into modern, universal terms. The local falls behind because it clings to the particular, and in particular to the web of signs signifying practices, customs, rites and institutions which cover over and conceal the continuum of natural forces. The local is a form of delay, or retardation that signifies too much. The loss of location is a consequence of radically enhanced mobility.

In turn, the hand-axe shrinks into a quasi-natural event in this picture. Even if it does constitute a 'vector' of technical progress or an advance in speed, it belongs in prehistory, where signs, consciousness, subjects, cultures and meaning are still only germinal. The neuronal account of the gestures attached to the hand-axe maintains continuity with a modernist account of the thermonuclear weapon to the extent that it describes hand-axes from the standpoint of hominid neuro-muscular parameters. (However, as we have seen, there are aporias in that standpoint that we need to discuss further.) If we accept that the hand-axe can also be understood as the (evolutionary) forces of nature at work in hominid corticalization, then it becomes an event described in universal terms. Again, local practices, signs, institutions and histories leave no trace. The hand-axe inscribes a limit to discourse concerning technical mediations because it is no longer clear whether anything technical (that is, anything that involves humans) is occurring. Myths of human origins must soon rush in to fill the gap if the threat to human uniqueness is to be met.

The nuclear weapon poses a limit because its speed and force cause us to ask: what kind of recognizably human individual or collective interest would want this degree of mobilization of energy? The hand-axe poses an inverted limit: what kind of recognizably human subject could not want or be capable of this? Each lies at the very perimeters of what we take to be human, individually or collectively. Each shades off into the non-human, the inhuman or dehumanized, the monstrous or the animal. The bomb represents the possibility of excessive force, and disregard for the consequences for humans and non-humans alike. The hand-axe represents something at the threshold of human intelligence. Other animals make and use tools, but these tools seem to be just on this side of the line between prehuman and human. Accepting a *radical* difference in speed between the bomb and the hand-axe entails accepting a profound continuity in relation to nature. One is the product of scientific manipulation of non-living natural forces, the other is simply living natural forces at work on the non-living. The technical mediations belong together because they are both reconfigurations of nature. Yet they lie apart at the opposite ends of history because one is energized by knowledge, subjectivity, representation and technology, and the other is mired in the prehistory of the human species.

The ending of this way of telling the story is not happy. From the other side of an irreversible break between the premodern and the modern, the premodern can only appear as what must be constantly left behind, or at least neutralized so that culture can become as universal as nature. Anything local must bend and collapse in front of the advancing shockwave of the modern. Making this split entails an attempt to coordinate a world according to one historical time, to render every place and idiom translatable into a universal de-Babelized tongue. The separation between modern and premodern carries within it the hierarchical privilege of the global – economical, political, epistemological or ontological – over the local, of the contemporary over the obsolete. Not only the past, but anything heterogeneous in the present becomes difficult to account for under this filtering regime. In the costly acceleration towards global power, signifying processes take on a new function: to mourn the loss of any proper place, any possibility of orientation. The earlier citation from Paul Virilio exemplifies this outcome. 'Mental confusion' regarding locality (temporal and spatial), and a collapse of the distinction between reality and fiction are sometimes celebrated, and sometimes lamented by humanists and posthumanists alike.

TOPOLOGY, MEDIATION AND MULTIPLICITY

This way of ending the story is dissatisfying only because it promises closure where there can be none. It dominated humanist and critical theorists' responses to modern technologies over the last century. It is possible to find versions of it in Weber, Heidegger, Adorno, Marcuse and Habermas. It purports to separate signification and technoscientific mediation from each other on the basis of a historical chasm between modern and premodern. Drawing on recent work in science and technology studies (particularly the work of Bruno Latour) and philosophy (especially Michel Serres), it is possible to roll the narrative backwards and look for branching points where a different kind of ending for the allegory might be opened.

Let us move back to the contrasts between the hand-axe and the bomb. Instead of seeing the differences between them as evidence of a radical historical break between premodern and modern technology which in turn maps back on to an ontologically loaded distinction between particular and universal, Latour and Serres would see both the hand-axe and the bomb as ways or detours along which collective life passes in order to stabilize itself. Latour writes:

> There is an extraordinary continuity, which historians and philo-sophers of technology have increasingly made legible, between nuclear plants, missile-guidance systems, computer-chip design, or subway automation and the ancient mixture of society, symbols, and matter that ethnographers have studied for generations in the cultures of New Guinea, Old England, or sixteenth century Bur-gundy. (1999, 195)

At the most general level, technologies such as a hand-axe or a thermonuclear device participate intimately in the articulation of humans and non-humans together in a *collective*. The two artefacts are not readily distinguishable on the basis that one belongs to a culture distancing itself from nature while the other belongs to a proto-culture mired in nature. Topologically, no such break exists. Rather, both entities hybridize human and non-human relations together in combina-tions that shift the limits of what a collective can do. Latour writes, 'by multiplying the hybrids, half object and half subject, that we call machines and facts, collectives have changed their topography' (1993, 117). Any technical mediation has a *specificity* which can account for its relative speed, but this specificity is not to be understood in terms of the separation between modern and premodern, or in terms of differences in kind. The effects of speed refer more to the differences in speed within and between collectives than to any absolute break. A way of grasping the *topological* differences between collectives is required. In crude terms, this approach asks how the hand-axe and the thermonuclear bomb *fold* bodies, living and non-living, together differently.

Instead of treating the contrasts in speed, complexity, durability and timing as given by the nature of the things themselves, the task is to sift out the interlacing and crossovers which underlie coarse distinctions between nature and culture, premodern and the modern, local and the global. Accounting for the specificity of any technical mediation involves reconfiguring what often appears to be a founding separation or gulf between these polar regions of modern versus premodern, or nature versus culture, as a set of unpredictably winding paths, passing through densely cross-hatched but under-represented equatorial continents of human/non-human collectives. The heavy historical oscillation between modern and premodern which sends signs and things either backwards into the premodern or forward into the modern (or postmodern) is deflected by an account of the topological complexity which technical mediations introduce into collectives.

FOLDING OF COLLECTIVE RELATIONS
THROUGH TECHNICAL MEDIATIONS

How is technical action involved in the constitution of collectives? In what sense is topology being used here? Two components figure centrally in Latour's account. First, he writes that 'the essence of a technique is the mediation of the relations between people on the one hand and things and animals on the other' (1995, 272). Note that the mediation concerns '*relations* between people' and 'things and animals'. Technical mediation does not directly link people and 'things and animals'. It modulates 'relations between people' through 'things and animals'. Second, if a divide must be mentioned, it should be the human–non-human divide rather than the premodern–modern divide. That latter divide makes it almost impossible to see how technology as a signifier can promise so much, and yet the entwining of technologies in collective life can remain under-represented. Technical mediations are inextricably involved in weaving together social relations within the collective at different rates and rhythms to produce the effects of speed. The topological complication arises because this weave diverts itself through things, animals, organisms. The collective – and in this context, 'collective' is preferable to 'society' or 'culture' – assembles humans and non-humans.

Mediation is only possible because of a margin of indeterminacy associated with humans. The idea that *as a multiplicity or collective, humans are more unstable than non-humans (living or non-living)* stands at the very centre of Latour's work on technical mediations. There are explicit formulations scattered throughout his writing: 'this is the big lesson of the philosophy of techniques: things are not stable, but people are much less stable' (1995, 277). Whatever technical mediations figure in our collectives, they respond to the greater degree of fluctuation, of sensitivity or instability associated with 'people', that is, with humans collectively.

If these instabilities propagated freely, human collectives would not survive. A *purely human* collective could eventuate but not endure. In other words, for Latour and for Serres, a purely human collective is unspeakable; there are only hybrid collectives woven together by perpetual displacement and slippage along various pathways mediated by non-humans. The only virtue attributable to technical mediation is relative stability, or in terms of originary technicity, iterability: technical mediations *slow* down the instabilities of collectives, absorbing, buffering, percolating or attenuating events, like a delta slows the flood of a river, or *speed* up instabilities, like levees on a river bank. At the core, they open different rhythms and rates of contact (faster and slower) within the multiplicity of elements that constitute a human collective. The notion of the speed of technical mediations shifts. Speed loses its absolute status and becomes relative to the fabric of a collective.

Certainly, humans often seem to be the most effective actors in a given grouping, but their agency and the consistency of the grouping they belong to is highly dependent on the way they mediate their relations through non-human entities. In Latour and Serres' terms, technical mediations are delegates or translators of the performance of human actors. The consistency of a collective stems from a set of relations. As Serres writes: 'The "we" is not a sum of "I"s, but a novelty produced by legacies, concessions, withdrawals, resignations, of the "I". The "we" is less a set of "I"s than the set of the sets of its transmissions' (1982, 228). A technical mediation is the name of the displacement, drift, delegation, delay or dispersal induced by the inevitable passage of relations between humans through non-humans, or 'quasi-objects', as Latour, following Serres, terms them.

FOLDING TIME

In the absence of any ontological division between the modern and the premodern, an explanatory burden falls more heavily on technical mediations. They articulate divergent realities together because of the detours, cross-overs and seams they introduce into the fabric of collectives. In the genesis of collectives, technical mediations channel instabilities into networks of non-human entities. They divert volatile, fluctuating relations between humans through alternate pathways, folding the collective by binding together different rates and rhythms. These diversions are not mere accidents. They are vital to the life of the collective. Without them, relations within the collective take on lethal instability. How can this interplay between stability and instability that occurs through technical mediation be accessed more precisely? As mentioned above, the break between premodern and modern is replaced by a topologically differentiated continuum. However, from Serres and Latour's perspective, temporality is enmeshed with topology. Latour, following Serres, says: 'Our first step is to look for the folding of time' (1994b, 45). Folding replaces radical separation. Every technical mediation brings together elements that are not strictly contemporary or simultaneous in terms of their genesis. As an assemblage or multiplicity, a technical mediation assembles heterogeneous elements from different times, from the paleolithic to the contemporary. 'Consider a late-model car,' Serres says. '[I]t is a disparate aggregate of scientific and technical solutions dating from different periods. One can date it component by component: this part was invented at the turn of the century, another, ten years ago, and Carnot's cycle is almost two hundred years old' (Serres and Latour, 1995, 45). From one perspective, this appears to be a trivial point. Of course the elements of an ensemble have differently dated origins. The thermonuclear bomb still involves technical elements such as bolts or screws; the invention of a bolt lies hundreds of years back in date from the vector processor of the supercomputer which handles the floating-point calculations needed to model what will happen to the bolt after detonation.[5] In that sense, there are clearly no purely present-day artefacts, only mixtures which associate elements inherited from different times. Furthermore, when we ask what heterochrony we can detect in the hand-axe, this 'dating' of components runs into a dead end. It does not have any elements. If time is folded through the technical mediation of a hand-axe, it does not involve differently dated technical elements.

However, the argument is more complicated than this. The apparent scarcity of technical elements in a hand-axe can be misleading. The different dates of the elements in a technical ensemble index the multiple paths and connections that compose the collective. Rather than unilinear progress from past to future, 'time flows in a turbulent and chaotic manner' (Serres and Latour, 1995, 45) because this mixing of technical elements introduces detours or transverse connections in the pathways within the collective. A collective or multiplicity entails networks of relations between the entities or actors that compose it. Technical mediations permit these networks of relations to vary their topology without always falling apart, or moving too far. When an alternate path between two points opens, the actors or nodes in those neighbourhoods are exposed to altered rhythms of contact. As two points in a network previously separated by a certain distance or delay become more closely linked, time within the collective folds in some way. Depending on the scale of the detour, the mesh of relations composing the collective has been augmented and distorted. When a different mediation enters the mesh of relations, different cycles and repetition strike up: thus time, the form of 'irreversible redundancy' according to Serres, is folded differently (Serres, 1995, 117). These rhythms, produced by synchronizations and delays, involve repetitions, cycles and feedback loops working against, with, through or independent of other repetitions, cycles, and beats.

SCALING IS FOLDING

How could such a move to remediate the divide between modern and premodern, or between nature and culture (to be premodern is to mix these two up too much, to be modern is to know the difference between them) account for the obvious differences in speed between the hand-axe and the nuclear bomb?

The variation between the prehistoric and the contemporary collective can be approached in terms of the *scaling* or *multiplication* in the number of non-human actors. This scaling up or mobilization enlists a greater numbers of actors, and it multiplies the paths that connect them, thus producing *networks* of technical action. The specificity of any given technical mediation is to be understood in terms of the relative scale of mobilization it introduces within the set of relations that compose the collective. The central argument of Latour's *We Have Never Been Modern* runs:

> The difference between an ancient or 'primitive' collective and a modern or 'advanced' one is not that the former manifests a rich mixture of social and technical culture while the latter exhibits a technology devoid of ties with the social order. The difference, rather, is that the latter translates, crosses over, enrolls, and mobilizes more elements, more intimately connected, with a more finely woven social fabric than the former does. The relation between the scale of collectives and the number of nonhumans enlisted in their midst is crucial. (1993, 109)

Any impression of a radical break between 'our' collectives and premodern or non-technological collectives is tacitly supported by the continuous extension and complication of our collectives through the scaling and mobilizing of non-human actors in networks.

The quantitative comparisons between the axe and the bomb corroborate this point: the difference between the two is essentially a matter of different rates, different timing or synchronization regimes generated by 'more elements, more intimately connected'. We can account for the increase in speed between the axe and the bomb in terms of the incorporation of a greater number of elements into the weave. But this accounting must also explain how the more extensive weaving together of elements is possible without assuming that it rests on an ontological dualism between rationality and nature, between mind and body. It is one thing to accept that contemporary collectives are intimately connected through large-scale technical ensembles of many different kinds. It is something else again to refrain from attributing the differences in scale between the hand-axe and the contemporary ensembles like the nuclear bomb to a different cognitive capacity to know and organize the world of things. The emphasis must therefore rest equally on the *number* and *intimate connection* (or 'folding') of the elements involved. The pathways which lead to the detonation of a thermonuclear weapon are instructive here. The speed of the exploding thermonuclear bomb derives from its design as a 'staged radiation implosion device'. A high explosive charge sets off a fission reaction whose energy is then focused on compressing the main hydrogen-deuterium fuel to critical mass. Detonation involves an energetically intensive interaction between 4000 or so components. We cannot account for the speed of the bomb relative to the hand-axe without considering the scale of interaction between different elements during detonation. That scale of interaction depends on careful coordination of the stages of the implosion. How can the movement of thousands of components in relation to each other be controlled or planned? In comparison to the people who made the hand-axes, nuclear weapons designers cannot simply 'shoot' their way to more efficient or explosive weapons (indeed they are even forbidden to do so since the comprehensive test ban treaty; MacKenzie, 1998, 109). Instead, they resort to massive computer simulations or 'codes' to stage the interaction of the 4000 components. Thermonuclear weapons from the outset were bound up with calculation and computation. The initial program run on ENIAC, the first American electronic digital computer, 'was a mathematical model of a hydrogen bomb for Los Alamos atomic weapons laboratories' (Edwards, 1996, 51). The escalation of force entailed in the bomb does not just reside within the 4000 components; it includes an associated milieu of calculation and information processing. The bomb could not be detonated without the simulations which render it workable. The speed of the bomb is linked to its simulation. In mid-2000, IBM announced the completion of the ASCI (Advanced Strategic Computing Initiative) White supercomputer. Under the terms of ASCI, computer companies build computers that can simulate the testing of nuclear weapons. The most powerful supercomputer in the world has been built specifically for the US Energy Department's Lawrence Livermore National Laboratory, a nuclear weapons facility (Hopper, 2000, 46).

Leaving aside all the delivery systems, the guided missiles, the naval and air arms, the communication and guidance systems that surround the bombs, the moment of detonation entails at least two kinds of mixing or interaction. These 'intimately connected' mixtures are clearly legible in the 'codes' or computer simulation that, as Donald MacKenzie points out, model the two main phases of detonation (MacKenzie, 1998, 109). The *mesh problem* belongs to the compressive and expansive phases of the detonation, during which the 4000 components are either imploding in order to produce the compressive force necessary to trigger the fusion reaction, or exploding after the chain reaction has finished. The *Monte Carlo* simulation belongs to the spasm of intense fluctuations that occurs in the nuclear chain reactions themselves.

The 'mesh problem'

Detonating a bomb is like making clay bricks in a mould. The mould figures as a set of 'frozen gestures' encountering the clay as it is tamped down into the mould by the hands of a worker. In the case of the bomb, the 4000 components making up the casing and primary charges briefly mould a fluctuating burst of high-energy radiation so that it can in turn form a compressive mould for the deuterium fuel. The difference here is that radiation can only form a compressive mould after a long detour through preparatory operations. Not only are 4000 components forming the mould, but this mould itself is prepared by the billions of computer calculations carried out in solving the mesh problem for a particular bomb design. The first main task of the designers is to focus the force of the exploding primary charges (which may themselves be smaller nuclear detonations) on compressing the deuterium charge to critical density. The process of accumulating greater density is crucial to the speed of any technical mediation. Without the compression produced by the implosive stage, a nuclear chain reaction won't take place. With it, the chain reaction becomes irresistably fast. Compression can only occur if the solutions to the mesh problem establish a configuration of components which contain the energy build-up. If too much energy flows out of the system, then the hydrogen which fuels the fusion reaction will not reach critical mass. This means both generating a 'shockwave' *and* retarding outward movement of the bomb's components until the main fusion chain reaction gets under way. The mesh problem 'involves modelling the evolution through time of a physical quantity or a set of interrelated quantities in a region of space. [A] numerical solution is attempted by superimposing a mesh of subdivisions in the relevant space . . . and calculating for a series of time steps the changing values of the physical quantities for all the points in the mesh (MacKenzie, 1998, 110). MacKenzie's study of the development of supercomputing indicates that the history of supercomputers can be seen, in part, as attached to the exponentially growing demands for more precise modelling of the dynamics of the interpenetrating fluid flows in nuclear weapons. In Latour's terms, we could view the supercomputer as an example of 'a slight enhancement of the mobility, stability and combinability of inscriptions' (Latour, 1987, 236). The enhancement in mobility of marks stands out clearly in the shift from the 'human computers' who carried out many of the calculations for early atomic weapons research and the supercomputers used by weapons designers in the late 1980s. As MacKenzie's comparisons show, the trend toward more precise modelling shifts the scale from a mesh of 2500 cells in the 1960s to a mesh of 125,000 cells in the 1990s, and in that sense, represents a direct scaling up of technical mediations (MacKenzie, 1998, 109). The network of calculations in fact prepares in advance and renders stable the interactions that occur during the technical mediations. The simulation slows down to around 10 hours an interaction that would otherwise take only a millionth of a second. The speed of the supercomputer and the massive multiplication of elements (many times 4000 components) accelerate the movement of marks in order to focus down the unstable implosion on to the deuterium fuel so that it can cross the threshold of critical mass.

The effectiveness of the implosion of parts is structurally coupled to an implosion of marks. The strong association between supercomputing and nuclear weapons design illustrates Latour's point about number and intimate connection. The scaling up of technical action that we find difficult to account for without attributing transcendental powers to the human or the technological in general, and to modernity in particular, eventuates as a topologically and temporally complex articulation of divergent realities. The supercomputer and the nuclear bomb are topologically linked in the sense that the iterative processes of calculating what happens in each cell of the mesh over time translates or maps an intensive energetic interaction on to an extensive, albeit compressed, surface of inscription. The marks moving in the supercomputer follow quite different pathways to the movement of components during the imploding stages of the thermonuclear bomb, but their trajectories lay down in advance the paths of the components during detonation.

In a certain sense, this is all obvious. Unless the bomb is modelled by calculation, it will misfire. The stronger point here is that the connection between the model and the mould is topologically continuous. The intensity of the bomb as a flash of energy and symbol of sovereign power cannot appear without the implosive mould whose shape and strength depend on the extended but compressed circulation of marks in the registers and memory banks of the supercomputers.

Monte Carlo simulation

Inside the 'mould' generated by implosion, a second computational task takes shape. The brief flash of radiation that comes out of the bomb follows a change of phase in the deuterium fuel after critical mass is reached. The situation is analogous to the moment when clay under pressure in a mould begins to redistribute itself as a brick. The computational task now concerns how what Simondon would call a 'pre-individuated meta-stable system' organizes itself within the limits temporarily imposed by the mould. The reorganization is called a 'chain reaction' to emphasize its multiplying or scaling effects. Under the pressure of implosion, a relatively small number of nuclear particles ally themselves to form new elements, throwing off radiation which triggers further alliances. Enabling this reorganization to continue calls for different kinds of computation to the mesh problem. The history of supercomputer architectures reflects these differences. A computer that performs well on the mesh problem may not perform well on Monte Carlo simulations. These simulations map what happens in the fusion reaction of a thermonuclear weapon by tracking the histories of a large number of particles (e.g. neutrons) as a set of branching paths. As J. von Neumann described it, the history of these neutrons and their progeny is determined by detailed calculations of the motions and collisions of these neutrons, randomly chosen variables being introduced at certain points in such a way as to represent the occurrence of various processes with the correct probabilities. (Hurd, 1985, 148)

If the mesh problem calls for an iterated solution of equations and leads to supercomputer computer architectures that accelerate floating-point calculations, the Monte Carlo method makes much less predictable demands, since '[u]p to 30% of the instructions in a Monte Carlo program may be branches' (MacKenzie, 1998, 111). The branches or conditional jumps in the Monte Carlo codes correspond in broad terms to the introduction of the 'randomly chosen variables' von Neumann refers to. (Again, as Donald MacKenzie's discussion shows, diverse supercomputer architectures responded to these computational demands. Speed in floating-point operations does not necessarily imply speed in conditional branch operations. However, the architecture of the Cray supercomputers developed during the 1960s and 1970s was able to accommodate both computational demands.) The Monte Carlo simulation draws up 'a map of the alliances and changes in alliance' (Latour, 1995, 277) that occur during the fusion chain reaction. In other words, if the problem of the *extension* of the networks of technical mediations is met with a mesh of calculations that prescribe the movement of technical elements towards a centre, the problem of intensity, of what happens at the centre, is met by the simulation of fluctuating interactions within a zone of singular intensity. Just as the preparation of the mould and the preparation of the clay converge in the making of a brick, two chains of technical mediations converge at the epicentre of the nuclear detonation. The transduction inherent to every technical mediation, no matter on what scale, is just that encounter between chains of different operations.

COMPRESSION, INFLECTION AND THE TOPOGRAPHY OF COLLECTIVES

The shockwave of radiation we see unfolding from 'ground zero' marks a significantly powerful reorganization of matter. If we look only at that flash, its force is blinding. The flash occurs because two long and winding chains of operations have prepared the ground for a specific, localized metastable state in an ensemble of elements. The two different computational tasks lie on opposite sides of this encounter, in much the same way that the abstract idea of a form and untouched raw material stand at the opposite ends of matter-taking-form. Without an extensive series of translations to bring them together, the mediation in which matter takes form cannot occur.

Compared to the explosion of the bomb itself, the computer simulations seem ontologically weak and insubstantial, even if they have strongly influenced the architecture and capabilities of contemporary information technologies. They only seem weak if the ensemble within which the detonation occurs is denied recognition. Without that ensemble, the few kilograms of hydrogen or helium that feeds the conflagration might as well be water or air. If, as Latour says, 'It is in the detours that we recognize a technological act . . . and it is in the number of detours that we recognize a project's degree of complexity' (Latour, 1996, 215), then the raw energy of the nuclear blast flows from a complex set of detours passing through supercomputers. If something happens at the epicentre, if there is in fact an epicentre, it can only be reached by passing along the winding paths which traverse numerous technical domains to finally reach the 4000 imploding components of the bomb, focusing compressive forces inwards on the slender supply of enriched hydrogen or helium fuel.

These detours weaving inscriptions, events and things together constitute a collective topography which existing notions of society find it difficult to grasp. What happens around the bomb I would argue is typical of many different technical mediations. The event of detonation cannot be separated from that topography. It grounds the event. The speed and force of the bomb gains traction in that domain. Detours institute shifts in the distribution of power and agency within collectives. What happens as the bomb detonates typifies technical mediations in their topological effects. They all stage just such a cumulative but staggered and detoured movement inwards towards a centre or relative interiority, whose reorganization triggers a move outwards, towards a newly defined periphery or outside. In the closed world of the Cold War, missile systems, communication and control systems, satellites and submarines encircle the globe, suspended in anticipation of that irreversible moment of detonation. But in preparing for it, something else happens. The technical ensemble that surrounds the bomb takes on its own instability. The force of nuclear weapons poses a limit to signification.

IS THE PRIMARY INSTABILITY HUMAN?

Latour and Serres look for *continuist* explanations of technical mediation. That is, they try to explain differences in speed, efficacy and complexity in terms of an uninterrupted gradient running between the two abstract poles of nature and society, or between premodern and modern. Their account cuts across the void that usually separates those poles. The preparation of the bomb shows that the continuum cannot be thought of as an homogeneous extension. The complexity of the bomb entails a multiplication of actors which do not lack complexities, depths and folds. It is topologically diverse. Movements of compression and inflection are multiple since, for instance, the bomb's force requires the supercomputer 'codes'. Our perception of technology can easily pass over these folds without noticing their depths. Changes in the topography of collectives derive from the mediation of social relations by non-human actors rather than large causes such as the universal power of modern scientific reason to control nature.

The unbridled scaling up of contemporary technology's power is due to 'our' almost complete incapacity to conceive of the depth of the mediation or hybridization that takes place through quasi-objects. It stems from the difficulty of conceptualizing the technicity of an ensemble as the effect of an articulation of diverse realities. Latour writes, 'the more we forbid ourselves to conceive of hybrids, the more possible their interbreeding becomes' (1993, 12). The instability and technological potency of contemporary collectives in contrast to the so-called 'primitives' stems from an unwillingness of the former to think through or represent the incessant detours and displacements passing through quasi-objects. As Latour writes:

> If . . . our Constitution [the collective schema that separates modern from premodern, society from nature, and human from non-human]

authorizes anything, it is surely the accelerated socialization of nonhumans, because it never allows them to appear as elements of 'real society'. . . . *The scope of the mobilization is directly proportional to the impossibility of directly conceiving its relations with the social order.* (original author's italics) (1993, 42–3)

The real instability or accelerating force resides in us, in the form of an interdiction: 'do not conceive of mediation or hybrids'. But because the whole tenor of Latour and Serres' argument resists any constitutive divide between human and non-human contingencies, and opposes any transposition of such a divide on to history, we have to be careful on this point. Is the 'modern' interdiction against thinking or representing technical mediation an historically contingent accident, or does it imply 'essential' instability? The form of this question should already be familiar. Again, it is the question of whether the urgency and importance that technology takes on in contemporary collectives can be deconstructed without discounting the originary technicity of those collectives.

The case of the hand-axe might allow this connection to be made more directly. Acheulean hand-axes were made for roughly one million years in a stable form (Ingold and Gibson, 1993, 227). Palaeontology calls them 'stereotypes' because of the constancy of their form. By contrast with almost any other human–technical mediation over the last 40,000 years, they constitute an apparently durable and stable form of technical action. If we hold to the tenet that humans collectively stabilize their relations through mediation, what kinds of instability are being folded, delegated, slowed down or made durable through the hand-axe? According to the archaeologists, innumerable so-called 'Acheulean hand-axes' are scattered over the three continents of Africa, Europe and Asia. But interpreting them in relation to any human collective is difficult. First, no one knows for sure what function the axes had or, indeed, whether they had any function at all. If they were technically functional, they could be the discarded cores from which other functional quasi-objects – flakes with sharp edges – were removed. This would mean they form a residue of a technical operation. They could be cutting tools – hence the name 'hand-axe'; they could be missiles like a discus, thrown at a herd gathered at the edge of the waterhole (many are found stuck in the mud around old waterholes) (Calvin, 1993). Second, although archaeologists have learnt how to make them, no one knows what it means for *homo erectus* to have made them. Their status as artifice or as artificial wavers. This status is precarious because their form, which seems so regular and deliberate to us, might not have been 'the intended shape of a tool' (Davidson and Noble, 1993, 365).

Their fabrication, function, significance and the collectives they were part of remain open questions. The prehistorical remoteness of the hand-axe constrains any empirical sequencing of the networks of associations through which a given technical mediation becomes active within a collective. We only have stones, bones and buried stores of fragments, to be archaeologically exhumed, sorted, classified and analysed. We lack the full context of possible utterances, rites, myths, gestures, movements, perceptions and points of view with which the hand-axe is articulated. If we cannot presume either *who* fabricated or used the hand-axe (if anyone), or *what* the hand-axe is (quite possibly not a hand-axe), we cannot place it in a network of associations and substitutions, whose series would delineate the individuation of a collective. In that case, questions such as these – what dynamics of the collective groupings are channelled through them? Are they slowing down or speeding up fluctuations in the collective, or both? – remain open.

Leaving aside all these difficulties of fabrication and function, but still assuming that some technical gesture of a living body is involved here, we can reconsider the basic problem of control of the hand-axe mentioned at the outset. The figures given then suggested that *if* the hand-axe (or the flakes derived from the stone) moved at speed, that is, if they were thrown or used to strike a blow, then a 'race condition' ensued. Assuming more or less stable timing constraints of human neurophysiology between now and then, a technical gesture involving the hand-axe could perhaps eventuate but not stabilize or be repeated. That is, its performance would not be repeatable, predictable, habituated or embodied. Because of the variation in firing times of neurones (11 ms) and the time of the feedback loops between peripheral muscles and the spinal cord (110 ms), the problem of controlling the trajectory of the throw or blow is just as acute and difficult as the problem of ensuring that a blast of a millionth of a second occurs in a predictable fashion. In both cases, the fluctuations of a single event must somehow be stabilized.

Like the simulation in advance of detonation through supercomputer calculations, the technicity of the hand-axes implies some form of anticipation, some planning in advance, some form of compression that precedes the movement away from the body that so many technical gestures involve (that is, the history of technology can be understood as the history of exteriorization of human life). The broad solution that human prehistory furnishes to this problem combines an upright posture, the 'freeing' of the hands from locomotion and, last, the 'freeing of the brain' or the corticalization or hypertrophy of the cerebral cortex of our anthropoid ancestors (Leroi-Gourhan, 1993). The whole organization of the 'anterior field of responsiveness' (to use Leroi-Gourhan's term) in humans points towards the embodiment of a capacity to anticipate.

Neurophysiological pictures of tool use (and associated language use) explain this anticipation in terms of networks. The race condition involved in throwing or hitting is overcome because action is mapped out in a network of cortical zones associated with hand movements. Action is repeated both synchronically and diachronically: first, there are multiple circuits of control in parallel which together average out the activation times, thereby improving accuracy; second, the control paths are trained and adjusted by earlier repetitions; third, and most important, the sequence of neuronal firing is 'buffered' or stored up in advance and then released at one go. We do not guide a technical gesture in 'real time' so to speak, but only through an accumulation of previous gestures, and through their repeated performance all at once (Calvin, 1993, 234–5). Pictured in this way, it is possible to regard the incorporation of a technical mediation as structured along the same lines as those we discussed in relation to the mesh problem and Monte Carlo simulations. Just as there is no possible detonation without a convergent series of detours, our bodies constitute a site of mediation – skeletally, neuronally, etc. – for the instability and fluctuating character of technical mediations. A gesture is the outcome of a living collective organizing itself around the mobility of a tool.

CORPOREAL INSTABILITY

If we take neurophysiological pictures of tool use at face value, they involve another kind of biological essentialism, this time centred on neurones and networks. This ending of the allegory risks becoming just as unsatisfactory as the modernist one, since now a biological or evolutionary straight line would be drawn between tool use and nuclear weapons. Technical mediations would involve no transductive interaction within collectives, because our capacity to use tools would already stem from biological attributes. Everything we do, including planning and building bombs, would result from that biological toolkit.

It is necessary to account for the neurological specificity of tool use without essentializing human tool use. Latour's approach is refreshing on this point. He asks, 'what then is a tool?' and answers 'the extension of social skills to nonhumans' (1999, 211). The hand-axe materializes through a movement in which relations between pre-humans shift across from a constantly fluctuating and decaying social order on to pliable but durable non-humans. What we regard as a tool is the outcome of a transductive negotiation between the transience and lability of social relations and the relative durability of living and non-living bodies: 'though composed only of interactions, the social realm becomes visible and attains through the enlistment of nonhumans – tools – some measure of durability' (p. 210). A tool is a detour taken by social relations.

Despite their apparently inert and static character, tools cannot be understood apart from living bodies associated in a collective. Susan Leigh Star and Karen Ruhleder write: 'A tool is not just a thing with pre-given attributes frozen in time – but a thing becomes a tool in practice, for someone, when connected to some particular activity. . . . The tool emerges in situ' (Star and Ruhleder, 1996, 112). Tools, too, materialize or corporealize transductively. If not, a variation of the old division between the social and the natural will be back in force, now in the guise of a split between social-technical layers and corporeal layers. A transductive event articulates divergent realities together to produce a number of different dimensions. During transduction, as we saw in the general case of matter-taking-form, pre-individuated potentials interact. What existed prior to the transduction as separate, although a convergent series, emerges articulated. We do not have to look for special neurological capacities to explain human tool use; rather we should look for ways in which neurological capacities attest to the incorporation of social relations via a detour called the Acheulean hand-axe (amongst others). During that detour, which lasts roughly one million years of prehistory, 'corticalization' occurs. This represents the corporeal dimension of the transductive articulation of social relations with stone. Like intricate links between supercomputers and the detonation of a bomb, we in our very bodies combine the compression and inflection of technical mediations, mobility and stability. Corporeally, we inherit a solution to the problem of speed of fluctuations.

SPEED OF THOUGHT

This ending of the story of how a hand-axe becomes a nuclear bomb comprises a different, still mythical account of human origins and their relation to technical mediation. It puts speed, urgency, fluctuations and instability in another light. Technical mediations take on a complex texture in their imbrication with collectives. In this alternative ending, the historical changes between the two technical mediations, especially in terms of speed, are neither transcendental, radical nor trivial. A convoluted, folded continuum lies between them which links bodies, collectives, technical ensembles and discourses. The beginning of this chapter suggested that corporeal theory, supplemented by science and technology studies, could provide traction on the difficult path away from a modernist take on technology in terms of speed and urgency. Specifically, I suggested that this alternative path would mean finding ways of affirming the genealogical depth of technical mediations within our collectives.

The work of Latour and Serres offers some elements of that affirmation. In their work, the associations of humans and non-humans are vital to collective life, not simply exterior to it. Association complicates social relations by sending or detouring them through relatively durable ensembles of non-humans. Technical mediation is a folding of social relations in order to render them durable. As modernists, we experience modern technology in terms of these attributes: unsurpassed speed, bewildering complexity and excessive power. The topological complexity of collectives cannot be mapped through the historical and ontological dualism of modern versus premodern, human versus non-human which that experience promotes. Radical divides validate the impression of unaccountable speed and complexity. In contrast, Latour and Serres account for the extension and power of contemporary technical mediations such as the bomb in terms of scaling (multiplication of actors) *and* topology. There is nothing absolutely new or surprising in the power of technology to change society, only the effects of a collective oversight in relation to its own embodiment of technical mediation. Relative to superpower collectives, nuclear bombs form a limit case. Relative to prehuman social groupings of hominids, stone axes form a limit.

A problem remains. The theory of technical mediation as social relations found durable answers to the need for a continuist account of how recent high technology seems to do so much in comparison to older or other technologies. But in the work I have discussed, the theory does not fully meet the challenge of corporeal theory. The continuist account substantively links social reality and non-humans without highlighting how social relations are also corporealized. Bringing the 'dark matter' of technical mediations within our collectives to light remains inadequate if the co-individuation of living bodies and technical mediations is still obscure. The case of the hand-axe made this apparent. The articulation of social relations through things changes what it is to have a body. The mixing is so thorough and deep that the folded surface of the cortex emerges as a consequence. Social relations are not only made durable in things. They are corporealized at the same time via the technical mediation. If a stone speeds up, if its flight time is reduced, what a body can do has changed; its limits have altered. At a very basic level, every technical mediation corporealizes a living body or bodies. It, to return to the vocabulary of corporeal theory, forms part of the iterative materialization of the surfaces, limits and matter of bodies. The transductive individuation of bodies in technical mediation enters into the divergent trajectories of living bodies in our collectives. Perhaps these transductive processes would allow us to think of ourselves as the 'weavers of morphisms' or 'exchangers of time'. If we weave, it is because we have been interwoven.

Conversely, the attention given to the topological complexity of our collectives offers something to corporeal theory. Corporeal theory, at least in the version I have been drawing on, has made strong use of the deconstructive notion of iterability. The power of institutional, cultural or political norms to regulate a body rests on the divergent iterability 'present' in and between bodies. Without that iterability, the norms would have no traction. But in the light of the folding together of humans and non-humans, the divergent iterability of living bodies can be rendered differently. The radical contingencies of embodiment (or in other words, the ways in which corporeality contest the prerogative of consciousness to give meaning to things) can begin to be articulated with technical mediations. Radical contingency need not be understood as a consequence of a specific indeterminacy in the human cortex nor as an essential human freedom. It resides in the ongoing co-individuation of bodies and things. Without this, the historical institution of our collectives would be nothing but a play on words.

NOTES

1. In his book *The Closed World: Computers and the Politics of Discourse in Cold War America*, Paul Edwards describes this scene in more detail (1996).
2. This is the major premise of Derrida's analysis of nuclear strategy (Derrida, 1984).
3. See MacKenzie (1998), Chapter 10, for an introductory account of thermonuclear reactions.
4. A 'race condition' is used in computer science to describe situations in which competition for access to some resource by two independently executing processes results in unpredictable, non-deterministic results. The interleaving of their operations is responsible for this indeterminacy.
5. On the role of vector processors in supercomputer architecture, see MacKenzie (1998, 102–3).

CHAPTER 3

The technicity of time:
1.00 oscillation/second to 9,192,631,770 Hz

[T]he simple pendulum does not naturally provide an accurate and equal measure of time since its wider motions are observed to be slower than its narrower motions. But by a geometrical method we have found a different and previously unknown way to suspend the pendulum; and we have discovered a line whose curvature is marvelously and quite rationally suited to give the required equality to the pendulum. After applying this line to clocks, we have found that their motion is so accurate and constant that, after many experiments on both land and sea, it is now obvious that they are very useful for investigations in astronomy and for the art of navigation. . . . Of interest to us is what we have called the power of this line to measure time, which we found not by expecting this but only by following in the footsteps of geometry.

Huygens, 1658

The second is the duration of 9,192,631,770 periods of the radiation corresponding to the transition between the two hyperfine levels of the ground state of the caesium-133 atom.

Blair, 1974

Three hundred years separate these two technical definitions of the second. The definitions both rely on oscillations of some kind – that of a pendulum, that of the radiation emitted by caesium atoms. Two different technological objects – a pendulum clock built for Christiaan Huygens in 1658 and an atomic clock dating from the 1950s, and forming the basis of current global timing standards – directly implement the definitions.[1] An infrastructure that now includes satellites stands behind the definitions. By virtue of the sheer multiplication of clocks in many different forms, clocktime may well be the most ubiquitous of modern technical infrastructures. If you had to choose a genuine universal associated with technology, clocktime would be a good candidate. Clocks are deeply embedded in diverse scientific, cultural, institutional, economic and military realities, and embodied in various 'body clocks' and 'clock chips'. Clock-faces are intimately woven into nearly every possible context. Through technological ensembles such as the Global Positioning System, clock-signals impalpably criss-cross every point on the earth's surface.

The contrast between these two different definitions of a second can be read as a symptom of the much broader perception concerning time and technology that has already been broached in the previous chapter: through technology, too much happens too fast. Speaking very broadly, technological speed can give the impression that the future is closed, and that any experience of time grounded in duration and memory has been lost. Living memory appears to be threatened by instantaneous retrievability. The complex rhythms of lived time are about to be overridden by the accelerating tempo of biotechnical interventions. Responding to this problem, there is even a well-publicized project in California, currently running under the name of 'The Clock of the Long Now', to slow down cultural change by building a clock that will run for 10,000 years and provide at least symbolic stability and longevity amidst the increasing ephemerality of globally resonant communication systems (Brand, 1999). The principal architects and sponsors of the project are computer engineers and artists, not critics of technology. The contrast between 1.0 oscillation/second and 9,192,631,770 Hz can easily be interpreted as a symptom of how human time is being lost to an inhuman, globalizing, technological 'time'. This time cannot be lived as such because its rhythms fall beneath the threshold of consciousness perception. Its measure is synchronized around the earth, and because technical performances are so often expressed in time-based terms (kilometres/hour, kiloherz, megaherz, gigaherz, milliseconds, microseconds, nanoseconds, 'teraf-lops', etc.), many contemporary technical mediations rely on an ever-present clocktime. Clocktime forms a basic infrastructural element of contemporary technology. Clocks time bodies, and machines, and mediate their linkages in diverse ways. The question is: how can the elementary technical mediation of clocktime become the object of a more differentiated and reflexive response? Chapter 2 broached the problem of separation between two limit artefacts (the hand-axe and the thermonuclear bomb). This chapter focuses on one highly concen-trated domain of technical mediation: clocktime over the last 350 years. It explores one possible response to the multiplication of oscillations by re-reading clocktime in terms of a concept of *technicity* drawn from the work of the French philosopher, Gilbert Simondon (Simondon, 1958/1989a, 1964/1989b, 1995).

The pendulum clock and the atomic clock are technical objects. More specifically, they embody technical mediations concerning the temporal and spatial sequencing of events. They stand between other entities, living and non-living, which together constitute a *collective*. As we have seen in earlier chapters, technical mediations knit social relations within human groups to non-living processes. In preliminary terms, 'technicity' refers to a specific virtuality or eventfulness associ-ated with this interweaving of living and non-living strands. At the risk of simplifying, it could be said that technicity is a term for the historical mode of existence of technical mediations. Technicity involves an event or inherently unstable genesis occurring at the limits of human–non-human collectives. As an evolutive or unfolding power, technicity structures particular technical objects, ensembles (such as the pendulum or atomic clock) and living bodies as provisional solutions to the problem of how an ensemble of living and non-living processes articulates its own temporal and topological limits. It precedes and overflows particular technological objects or social func-tions. Particular technological objects or technological ensembles are, in Simondon's terms, 'objectivations of technicity' (Simondon, 1958/1989a, 163).

To speak of the *technicity* of time implies an evolutive, genetic process subject to objectivation (for instance, in different clocks) and 'subjectification' (for instance, in various embodiments of clocktime), yet remaining irreducible to particular technical treatments. Conversely, to focus on the technicity of *clocktime* is to suggest that the social ordering of time, the scientific manipulation of time as a fundamental dimension of physical systems, and the individual experience of time as imbricated in memory and anticipation, all pass through *metastable* processes associated with clocktime as an event.[2] Such an event may well escape historical dating practices or a technological history of clocks. Those practices and knowledges must, in order to begin their work, rely on a certain ensemble of other technical mediations which are neither neutral nor transparent in relation to time. *Clocktime technicity* runs through the two exemplary objectivations of time under discussion: Huygen's pendulum clock and the atomic clocks currently used, for instance, in the Global Positioning System (GPS) constructed at the orders of the US Department of Defense. The first main section of this chapter highlights two more or less conventional accounts of clocktime which regard it as a symptom of the capture of time by modern social and technoscientific processes of ordering events. The second section sketches a more nuanced approach to clocktime, taking into account both the technical specificities of timing regimes and the divergent realities which unfold out of clocktime. The final section develops this approach into an explicit account of the technicity of clocktime as an ongoing differentiation or genesis occurring at the limits of sociotechnical collectives.

MODERN TECHNOLOGY AND THE LOSS OF TIME

A symbolic connection runs between Huygens' pendulum clock and the dozens of atomic clocks installed in the GPS. On the one hand, the pendulum clock offers an elementary symbol of incipient technological globalization, while on the other, the atomic clock symbolizes the completion of a certain kind of geotechnical globalization. Already in Huygens' book of 1673, where he sets out a geometrical deduction of the isochronism of the pendulum's oscillations, the importance of clocks to sovereignty, to navigation and astronomy is at the forefront. In dedicating *The Pendulum Clock, or Geometrical Demonstrations Concerning the Motion of Pendula as Applied to Clocks*, to Louis XIV, Huygens writes:

[F]or since my clocks were judged worthy to be placed in the private chambers of your palace, you are aware from daily experience how much better they are in displaying equal hours than other such instruments. Further you are not unaware of the more specialised uses which I intended for them from the beginning. For example, they are especially well suited for celestial observations and for measuring the longitudes of various locations by navigators. (1986, 8)

The pendulum clock links the private chambers of the king to the peripheries of empire and colonies. Otto Mayr writes: 'Princes and their courts led the way. Courts, with their complex ceremonials and their many-layered staff hierarchies demanded punctuality. . . . The clock became an attribute of nobility' (Mayr, 1986, 16–17). It secures the possibility of moving between the centre and periphery, by capturing invariant oscillations to mark the time and providing a technical realization of a universal, coordinated time and length. The pendulum clock modulates the incipient movement of globalization by establishing the possibility of a globally valid measurement system. The atomic clock, by contrast, affords an image of the completion of globalization, at least in relation to navigation, in the guise of the Global Positioning System (GPS) which relies on many atomic clocks located in orbiting satellites and in a network of ground stations. As Paul Virilio wrote:

A new type of watch has been on the market for a while now in the United States. The watch does not tell you the *time*; it tells you *where* you are. Called the GPS – an abbreviation for *Global Positioning System* – this little everyday object probably constitutes the event of the decade as far as globalization of location goes. (1995a, 155)

It is not clear whether GPS did constitute the event of the decade. A recent article on GPS remarks that 'the history of GPS is a classic case of a technology in search of a market' (Tristram, 1999, 70). On the other hand, the Presidential Directive on GPS signed into law by President Clinton in 1996 guarantees some ongoing life for GPS.

Nonetheless, GPS, with all the clocks it puts into orbit, aptly symbolizes the 'globalization of location' referred to by Virilio. Its exactitude now reveals unpredictable variations in the revolutions of the earth itself. Many different narratives of progress, technological evolution, alienation or epochal shifts could form the backdrop to this perception of the globalization of clocktime, and its suppression of different perceptions or experiences of time. In broad terms, Heidegger's view expresses a typical objection that the twentieth century of critical theory and philosophy had to clocktime:

> But time cannot be found anywhere in the watch that indicates time, neither on the dial nor in the mechanism, nor can it be found in modern technological chronometers. The assertion forces itself upon us: the more technological – the more exact and informative – the chronometer, the less occasion to give thought first of all to time's peculiar character. (1972, 12)

Consistently through his work, Heidegger insists that clocks provide no insight into time.[3] The reasons for this are complex and require extended discussion. In summary, Heidegger provides an extremely powerful account of the ways in which time marks a limit for thought. While the links he makes between temporality, being, thought and history shift somewhat in the course of his work, it is always the exteriorization or objectification of temporality as time that he questions. Clocktime figures as a central instance of that objectification. It is coupled to modern science and the subject–object distinction. Clocktime glosses over the unstable interplay of being. In his terms, we cannot speak of time itself on the one hand and the modalities of consciousness apprehending the flux of time on the other without lapsing back into a metaphysics of subjectivity. Only a single process of temporalization occurs. It is composed of complex movements of anticipation and repetition through which provisional forms of stability, practice, institution, subjectivity, memory and historical existence unfold. Against this sophisticated framework, the core objection to clocks is simple. Through clocks and clocktime, attention shifts away from the coalescent and emergent character of temporality. Not just clocks, but modern technical mediations more generally, deflect attention from the deep and intricate intermeshing of time, bodies and thought. Through modern technologies of timing, profound differences in history, tradition and hence cultural existence are lost.

CLOCKTIME AS SOCIAL INVENTION?

As an alternative to this critical repudiation of the technical mediations of time, we could, for instance, understand clocktime as a social event or invention. This is the thesis of Norbert Elias, when he writes:

> By the use of a clock, a group of people, in a sense, transmits a message to each of its individual members. The physical device is so arranged that it can function as a transmitter of messages and thereby as a means of regulating behaviour within a group. (1993, 15)

Clocks from this perspective are 'simply mechanical movements of a specific type, employed by people for their own ends' (p. 118).[4] The notion of time as a 'relatively more unitary human-centred concept' (p. 115) effectively presents clocktime as a sociosymbolic invention concerned with more precisely regulating and coordinating the repetition of social phenomena.[5] In Elias' rich account, the problem of how technology accelerates the experience of time can only be understood in terms of the 'specific capacity of people for envisaging together and, thus, for connecting to each other what happens "earlier" and what happens "later", what "before" and what "now" in a sequence of events' (p. 74). The 'sociocentric' nature of time consists of the use of repetitive, usually inorganic sequences (ranging from the movement of a shadow during the course of the day to the hands or digits of a clock) for the symbolic representation of non-repeatable social sequences. The very existence of time, by this account, is a social artefact of the numbering or ordering of sequences in synchronized relation to each other, and this in turn relies on physical phenomena that display numerable repetition. Timing technologies display sequences of marks through which social groups code sequences of events. So, according to Elias, looking at a watch:

[W]e read and experience the changing configuration of these moving units on the face of a watch in terms such as 'five minutes past seven' or 'ten minutes and thirty-five seconds'. In that way, moving configurations of marks used for timing events are transformed by social customs of the beholders into symbols of instances in the flux of incorporeal 'time' which, according to a common use of the term, appears to run its course independently of both any physical movement and any human beholder. (p. 120)

The 'social customs of the beholders' as well as the 'specific capacity' of people to envisage sequences of events are responsible for the emergence of autonomous physical time. Time *appears* to have an existence independent of social customs, but in reality it is grounded on the social acts of timing or dating by which phenomena are semiotically related. The only problem with what Elias suggests here is the nature of the transformation between different social timing regimes. While he says that 'the significance of this emergence [of autonomous clocktime] can hardly be overrated' (p. 115), the specificity of the artefact which connects events in a sequence (the clock) does not figure either in the social customs or in the specific capacity to envisage (i.e. remember) temporal orderings. The lines between bodies, things, society and nature are secured.

CLOCKTIME AS TECHNICAL MEDIATION

In such a summary account of two quite different perspectives on clocktime, stereotypes are hard to avoid. Despite that, Heidegger's and Elias' approaches have affinities. Neither have any way of affirming the interval between 1 oscillation/second and 9 billion oscillations/second. Either it must be regarded as destructively superficial in relation to the real problem of how to think (Heidegger), or it must be accepted as deriving from social arrangements which, so to speak, are 'more real' (Elias). To speak of the technicity of clocktime is to try to offer some way of comprehending the multiplication that has occurred between 1 oscillation/second and roughly 9 billion oscillations/second without repudiating the role of technical mediations.

In one important respect, a lead on the technicity of clocktime comes from recent work on the role of technical mediations in the formation of collectives. As we saw in Chapter 2, accounts of technical mediation such as Bruno Latour's *We Have Never Been Modern* argue that the broadly modern habit of thinking in terms of a radical break between the social and the natural (or between subject and object, sign and thing, . . .) tacitly permits technical mediations to multiply. Because technical mediations cannot be directly figured as integral to the life of the collective, because they are always seen as mere supplements to a core social reality, the collective work that they represent constantly recedes from view. We have so much 'technology', comparatively speaking, because the collective is trying to catch its own tail. From Latour's perspective, 'if . . . our [modern] constitution authorizes anything, it is surely the accelerated socialization of [technical] nonhumans, because it never allows them to appear as elements of "real society"' (1993, 42). Certainly the representative positions I have sketched around Heidegger's and Elias' work conform to this judgement. In both Heidegger's 'the more exact the clock, the less occasion to give thought to time', and Elias' 'we read the clock as symbols of instances in the flux of incorporeal time', clocktime always remains secondary to a more primary ontological or social reality. If clocktime is always seen as a deadening of lived time, then all the technical objects and ensembles which structure themselves around clocktime risk being excluded at the same time. Given the pervasiveness of clocktime, this outcome constitutes a general disengagement with technology.

Bringing a transductive approach to a specific case such as clocktime entails apprehending clocks as a technical mediation which does not measure or administer a pregiven social *or natural* time or space, but which constitutes a regime of timings and spacings from which society and nature, time and space unfold. To be sure, clocks are not alone in defining time. Every technical mediation, insofar as it folds, deforms and shifts relations between living and non-living elements of a sociotechnical ensemble, 'eventalizes' times and spaces. In this emergentist perspective, the 'time' of a medieval castle and the 'time' of a high-speed train are not the same because of the different topological and temporal folds they weave into collectives (Latour, 1997, 179). The specificity of clocks and clocktime consists only in their special status as guarantors of a certain regime of homogeneous time (iso-chrony) and homogeneous space (isotopy). Subsequent bifurcations into objective and subjective times only gain traction through the incremental synchronization of more and more clocks. To quote Latour:

> This does not mean that we are *in* an isotopic space and an isochronic time, but that locally, *inside* metrological chains, there are *effects* of isochrony and isotopy produced by the carefully monitored and heavily institutionalised circulation of objects that remain relatively untransformed through transportation . . . rods, hands of clocks, gears and structural isomorphies. (1997, 185)

In other words, there is no space and time apart from the technical mediations through which selected events — oscillations and inscriptions, in the case of clocktime — are linked. (The very axes of synchrony and diachrony which have organized wide domains of recent critical thought could appear quite different if we took this perspective seriously enough.) The specificity of clocks as 'metrological chains' undergoes reconfiguration here since they now function as what Simondon calls 'key-points' [*point-clefs*] from which times and spaces unfold.

An example of this analysis in action can be found in Bowker (1995). He states the problem succinctly: 'It is unproblematic to say that societies with differing configurations of economic and technolo-gical development have differing ways of understanding and represent-ing time and space. The problem begins when one tries to move out from this statement in any direction' (p. 47).

Through his discussion of nineteenth-century technological infra-structures such as railways and factories (technical mediations that lie within the interval under discussion here), Bowker accounts for the *convergence* of a neutral isotropic space–time compounded with large-scale patterns of social organization. The crucial implication for our purposes would be that the interval between 1 oscillation/second and 9 billion oscillations/second stems from the way that infrastructural technologies such as railways are constantly 'conjuring nature' into a particular representational framework in which time and space are universal and neutral (Bowker, 1995, 63).

Nonetheless, there is at least one point at which it may be necessary to diverge from this line of thought. As yet, it has been difficult for this fruitful and consequential body of work to affirm ongoing dynamism or instability in sociotechnical collectives other than as a consequence of our own inability or unwillingness to explicitly think and represent technical mediations (hence Latour's semi-ironic proposal for a 'Parliament of Things' (Latour, 1993, 142–5)). Latour, for instance, says that once we recognize the constitutive role played by technical mediations, we will then be able to 'sort times' (Latour, 1993, 76). This suggests that, despite the mutability of almost every other relation or entity, times could be taken as remaining stable.[6] If times (in the sense of the ordering of series) remain stable, proper representation could put the sociotechnical ensembles in which we live back on a stable footing. Ultimately, instability is either stabilized, or ephemeral.

The critical and sociocentric accounts of clocktime view the interval between the pendulum clock and the atomic clock as stemming from either (a) a general loss of time attributable to modern technology; or (b) a particular social habit of overlooking how social relations symbolically order events. While Latour's approach departs radically from these two accounts, it tends to say that the effects of technological speed result from a failure to properly represent the role of non-human technical mediations in stabilizing ephemeral events as essences (i.e. 'the time') within sociotechnical collectives. Despite crucial differences from Heidegger and Elias, the interval between 1.0 oscillation/second and 9,192,631,770 Hz still derives from a failure to think, and instability is always transient, not ongoing. While Latour may be compelling in his judgement of the modernity's inability to represent its own deeply technical mediated constitution, it would be somewhat more 'enabling' if the under-representation of the technical mediation of time was not solely attributed to a failure, and if instability was not always secondary to stability. Sometimes it seems that behind this assessment, there is an implicit promise that stability is the norm and instability the deviation. Perhaps an account that emphasizes structural instability or metastability would also be of interest.

TECHNICITY AND ISOCHRONISM:
MECHANISM AND GEOGRAPHY

Clocktime hardly seems a promising place to envisage ongoing instability or metastability since it epitomizes regularity and stability. The growth of this regularity, stability and autonomy, I will argue, offers a key instance of the oscillatory or modulating aspect of technical mediations when they are understood as emerging from an articulation of diverse realities. Following and extending Latour's account of a technical mediation as an *event* might bring us closer to the multiplicatory rhythms of clocktime as evolutive power of divergence. Clocktime's mode of existence as technicity is purely neither social nor technical. In an article entitled 'Time and Representation', Isabelle Stengers and Didier Gille explain how time can appear to be or become autonomous through clocks (Stengers and Gille, 1997). They too speak of the division between social and natural/scientific times, but they discuss how the split between social and scientific times eventuates through the specificity of a particular technical artefact, the pendulum clock. Their complex account of the relation between time and the technical specificity of clocks introduces considerations that will be useful in thinking of the evolutive or divergent character of the technicity of clocktime.

MECHANICAL ISOCHRONISM

Stengers and Gille describe an historical process out of which the social, scientific, natural and technical phenomena of timing unfold: 'the concrete object whose introduction marks the establishment of an autonomous law of time can be more precisely identified with the pendulum clock that Christiaan Huygens constructed in 1658' (Stengers and Gille, 1997, 183). Natural time and social time are only the abstract poles of a zone of interaction which can be optimally read in terms of the technical specificity of the pendulum clock. Huygens' pendulum provides a point of inflection or 'intrinsic singularity' (Deleuze, 1993, 15) for the technicity of clocktime. Stengers and Gille write:

It is usually claimed that Galileo's discovery of the law of pendular motion [the correlation between the length of a pendulum and the period of its oscillation] at last gave a scientific solution to the technical problem of the measurement of time. . . . However, Galileo did not produce such a mechanism: the free pendulum is a pure phenomenon; the oscillations need to be counted and the movement periodically restarted. (Stengers and Gilles, 1997, 185)

The passage cited at the outset from Huygens' *Pendulum Clock* verifies the last point. Huygens writes, 'the simple pendulum does not provide an accurate and equal measure of time. . . . [W]e have found a different and previously unknown way to suspend the pendulum' (1986, 11). While his way of suspending the pendulum is not absolutely singular, the pendulum clock possesses a technical specificity that distinguishes it from previous time-measuring techniques such as the foliot clock found in medieval clock towers. Huygens points to this when he declares that he has unexpectedly found a different way to *suspend* the pendulum. Mechanically, 'the foliot clock . . . appeared as a complex in which everything participated in the definition of the speed of the clock hands, without it being possible to specifically identify one element as regulator' (Stengers and Gille, 1997, 184). With the pendulum clock, 'the work of the clockmakers will largely consist of disconnecting, as much as possible, the pendulum-regulator from the rest of the mechanism' (p. 186). The decoupling of the pendulum from the rest of the clockwork takes various forms – recoil escapement, deadbeat escapement, free escapement, constant force escapement – yet all these forms head in the direction of presenting the isochronic oscillations of the pendulum as an embodiment of 'the time'. The remainder of the mechanism becomes a means of either displaying information about the time or correcting for the fact that the pendulum itself is never ideal, that it always suffers from friction, and that, more importantly, as Huygens points out, the period of a simple pendulum varies according to the driving force of the clockwork. If the pendulum can be isolated from these variations, then time itself can appear to be separate from its technical realization. Through isochronic oscillation, the pendulum can exist as the autonomous embodiment of natural or physical time.[7]

GEOGRAPHICAL ISOTOPISM

The series of escapements tends to isolate the pendulum from all variations, so that it becomes a pure source of information for the work carried out by the clock in moving its hands. But this isolation only makes sense if the motions of the pendulum have more than local value. While, as Galileo recognized, the motions of the pendulum are more or less isochronic so that they can help establish proportional relations between speeds, they remain localized if their period is not standardized. Thus, write Stengers and Gille, 'the [earlier] pendulum clock developed by Huygens in 1657 had a period of oscillation of 0.743 seconds. This number had no *raison d'être* other than it corresponds to a particular set of cogwheels' (p. 190). The metrological chain of clocktime can only be extended if the oscillations of the pendulum can be tied to something else. Huygens, developing the relationship between pendulum length and the period of the pendulum, in 1658 constructed a clock that beats once per second. The geometrical (and hence heavily *diagrammatic*) proof of the isochronic oscillation of the pendulum of this clock was published as *The Pendulum Clock* in 1673. Furthermore, it includes instructions on how to standardize the hours measured by the clock to the revolution of the earth on its axis.[8] The oscillations of the pendulum become geometrically and geographically isochronic; thereafter, movement of the cogwheels is subordinate to the length of the pendulum, itself calibrated by the regularity of the earth's revolutions. No longer expressing the relative speeds of phenomena, the time of the pendulum clock becomes autonomous or, at least, it will be represented as being autonomous. The constitution of the second as a unit of time coupled to the earth's revolutions allows it to claim independence from all terrestrial locality. It is now identified with the earth's diurnal revolution rather than the alternation of day and night, or the apparent movement of the stars, which vary seasonally and from place to place. Whereas the foliot clock of the medieval clock tower proclaimed variable hours adjusted to fit the varying length of the solar day at particular places, the pendulum clock measuring the standard second displays constant hours, regardless of the time of year or the location of the clock. As Stengers and Gille observe, 'objective, regular, normalized time, existing by and for itself, is born, uncoupled from what is now no more than the straightjacket of phenomena' (p. 191). The cost of normalized time is a strengthened relationship between the pendulum's and the earth's movements.

At the end of *The Pendulum Clock*, Huygens goes on to propose that a global standard of measurement should be derived from the pendulum clock. Again the connection between timing and spacing is unavoidable, and the pendulum clock ends up looking like a precursor of the global clock system that GPS implements:

> A certain and permanent measure of magnitudes, which is not subject to chance modifications and which cannot be abolished, corrupted, or damaged by the passage of time, is a most useful thing which many have sought for a long time. . . . [T]his measure is easily established by means of our clock, without which this either could not be done or else could be done only with great difficulty. (Huygens, 1986, 167)

The method that Huygens offers as the basis of a universal, atemporal standard of length involves tuning a simple unregulated pendulum's oscillations to the regulated oscillations of the pendulum clock by adjusting its length. The length of the synchronized simple pendulum will be the universal 'hour-foot', a length that will be the same at all places which share the time of the pendulum clock. Such a measure would have been valid everywhere and 'for ages to come' as Huygens hoped, if there were not variations in gravity at different points on the earth's surface.

This proposal for a universal measure opens a continuous path from the oscillations of the pendulum clock to the oscillations of the atomic clocks used in GPS, even if it is convoluted and full of fluctuations. In 'the ages to come' mentioned by Huygens, when for instance the measurement of distances is carried out through GPS, a local unregulated oscillator in a receiver (a quartz clock) is tuned to the regulated oscillators of the atomic clocks in at least three GPS satellites. Synchronization is now no longer carried out by hand and eyes, but via a set of circuits in the receiver which modulate the local oscillator of the GPS receiver until it coincides with the oscillations being transmitted by the GPS satellites (Kaplan, 1996, 121). The location of the receiver is determined by finding the intersection of the distances from those three satellites. The oscillations stretch between the pendulum clock and GPS: the speeding up between 1.0 and 9,192,631,770 oscillations/second requires that one oscillator be tuned to another oscillator of known period in order to synchronize the oscillations, to allow them to resonate with each other. Perhaps, one important aspect of what we experience as globalization today is the cumulative effect of the synchronization of dispersed oscillations.[9] Heightening the suspension even further, the atomic clocks orbiting in the GPS satellites are themselves synchronized ultimately to another global timing standard: UTC, Universal Coordinated Time. UTC exists nowhere as such; there is no single master clock. UTC is a 'paper standard', with no concrete embodiment apart from the statistical procedures used to correlate and synchronize several dozen atomic clocks scattered around the globe (Kaplan, 1996, 55).

In broad terms, the technical specificity of the pendulum clock begins to reveal some singularities that the other accounts of modern technology and time had to varying degrees denied it. Stengers and Gilles' argument does not present the pendulum clock as determining what time, in the form of clocktime, is. Rather, it seeks to show how the emergence of autonomous clocktime, or the 'physical time' which Elias described as 'branching off' from social time, requires a specific and localized decoupling of pendulum and clockwork together with a specific yet generalized coupling of the pendulum with an associated milieu, the revolutions of the earth and its gravitational field. Through a pendulum's suspension, and the resulting mechanical isochronism of its oscillations, clocktime can appear as the embodiment of autonomous time. Through synchronization of oscillating pendulums, the pendulum's oscillations geographically distribute clocktime and serve as a way of establishing spatial relations between distant places.

METASTABILITY: FROM ISOCHRONISM TO EVENT

In terms of the three different perspectives on clocktime discussed above, we could ask: is this tuning or resonance between dispersed oscillators best understood as the homogeneous extension of the essence of modern technology, as the transmission of a message-regulating behaviour within a social grouping, or as the institution of an 'immutable mobile', an entity which can maintain some form of constancy as it is translated across different contexts (Latour, 1997, 180)? It might be something of all these. All three emphasize stability and homogeneity within an expanding collective. Later technical developments of the pendulum clock, especially during the eighteenth and nineteenth centuries, concentrate on refining the isochronism of the pendulum over a wider range of milieus, taking into account variations in air pressure and temperature (Howse, 1980). When the oscillator becomes piezo-electric, as in the case of the quartz crystal oscillator in the early twentieth century, or atomic as in the GPS of the late 1980s, this is partly an attempt to maintain isochronism over an ever wider range of milieus. The technical problem remains constant throughout: how can the isochronic constraint necessary for autonomous time be maintained over a wider range of milieus? From the perspective of the technical history of clocks, it can only be maintained if the clock can stabilize itself in the face of new sources of variation.

However, from the perspective of technicity, and for an understanding of the divergence symbolized by the interval between 1.0 and 9,192,631,770 oscillations per second, *metastability*, rather than *stability*, is crucial. So far, this account of the technicity of clocktime has mainly concerned the *stabilization* of an autonomous time through isochronic oscillation. I want to focus now on an ongoing *metastability* associated with clocktime. The emphasis that Huygens himself places on the *suspension* of the pendulum and, in particular, on the precise geometrical description of the two curved plates which limit the motion of the pendulum, can be understood from the perspective of metastability.

Metastability refers to the provisional equilibrium established when a system rich in potential differences resolves inherent incompatibilities by restructuring itself topologically and temporally (Simondon, 1958/1989a, 154–5). In Simondon's preferred example of a physical metastability, a super-saturated chemical solution begins to crystallize. As it does, it 'individuates': some singular point – an impurity, a seed crystal – in the solution permits the solution to restructure itself as a growing crystal. The crystal structures the energetic potentials of the solution (Simondon, 1964/1995, 24). At the point of crystallization, the solution is metastable. The growth of the crystal represents a provisional resolution of the potential differences that precede it.

There are significant differences between a super-saturated solution and a sociotechnical collective. Yet considering the fact that the collectives we are talking about extensively couple living and non-living processes, there are points of contact here. Clocktime as it moves between 1 oscillation and 9 billion oscillations per second can be seen as a temporal and topological ordering that continues to unfold from a metastability. The way in which clocktime incorporates new sources of variation, and restructures itself in the process, can be compared to the provisional resolution that a crystal represents for the metastable super-saturated solution. The question is: what metastability are we talking about here? What *virtuality* (to use Latour's term (Latour, 1997, 190), who draws it from the work of Gilles Deleuze (1993), who in turn perhaps borrowed it from Simondon) inhabits the accelerating trajectory of clocktime?

By framing metastability in terms of our own collective involvement with things, it might be possible to preserve certain valuable elements of the three approaches represented by Heidegger, Elias and Latour, and to add something worthwhile. From Heidegger comes an insistence on temporality as concomitant of any attempt to think what is. Temporality allows Heidegger to think radical finitude; that is, as historical existence in the absence of absolute foundations. In Elias' work, the social or group character of timing finds expression. His work alludes to the historically variable constitution of timing regimes, and their diverse social functions within social groups. Latour provides something different again: the 'dark matter' of the societies we inhabit consists of the manifold technical mediations that stabilize the collective so that something like history and time becomes possible. We might now be in a position to add something at the intersection of these very different approaches: the metastability of our collective involvement with clocktime.

LOCALIZATION OF INDETERMINACY
IN CRITICAL PHASES

Let us locate this metastability more specifically. No matter how ideal the motion of the pendulum becomes over a range of conditions, it still involves contact in which energy of some kind is converted into information displayed on the face of the clock. There are always moments of contact between the pendulum and the clockwork it regulates. As Stengers and Gille suggest, the technical development of the pendulum and escapement tends to minimize the energy converted in this process, and to constrain it as a one-way process in which information flows from the pendulum regulator to the clock-face, but not the reverse. From the technical perspective, contact between the pendulum and the clockwork represents a deviation away from the ideal of the autonomous pendulum. Perfect suspension as an ideal seeks to disguise the technical constitution of clocktime. If an irreversible expenditure of energy appears in the constitution of clocktime as pure information, then the autonomous character of time would be under threat.[10]

From the standpoint of the technicity of clocktime, however, this transfer of energy is critical: the intermittent moments of contact constitute the metastability of the system. In effect, these moments constitute clocktime. Simondon describes the technicity of machines in general as a capacity to be repeatedly informed through a carefully staged ensemble of critical phases:

> [T]he existence of a margin of indeterminacy in machines should be understood as the existence of a certain number of critical phases in its functioning; a machine which can receive information temporally localises its indetermination in sensible instants, rich in possibilities. . . . Machines which can receive information are those which localise their indetermination. (1958/1989a, 141)

What was earlier termed the 'decoupling' of pendulum and clockwork can now be described somewhat differently. In Huygens' clock-machine of 1658, the pendulum repeatedly comes into contact with a rod attached to the clockwork escapement: '[T]he small rod . . . , which is moved very slightly by the force of the [clockwork] wheels, not only follows the pendulum which moves it, but also helps its motion for a short time during each swing of the pendulum' (Huygens, 1986, 16). Contact between the rod and the pendulum localizes indetermination in a very specific way. The pendulum, whose associated energetic milieu is the earth's gravitational field, encounters the cogwheel gear-train of the clockwork, whose associated milieu encompasses the rituals and protocols of the king's bed-chamber, where Huygen's clock stands. Despite worn or sticking cogs, the pendulum 'informs' the rod of the period of its oscillations. During 'the short time', the pendulum enters into a complicated and highly mediated exchange with the potential energy stored in the weights that drive the clock. The gravity-driven clockwork transfers some of its stored energy to the pendulum and, reciprocally, the pendulum's oscillations inflect the rhythm of the clockwork's movement. The 'sensible instants, rich in possibilities' that Simondon refers to occur during the wavering, inconstant contact between rod and pendulum. Out of the super-saturated, undifferentiated potentials of those instants, two divergent realities unfold, one facing towards a geographical-terrestrial milieu (the earth's gravitational field), the other facing towards a social milieu of symbols, numbers and counting conveyed as 'the time'. As Huygens says, 'it will always measure the correct time, or else it will measure nothing at all' (p. 16). To measure correct time is to maintain the interaction between these two milieu, and prevent their incompabilities from becoming too great.

The atomic clocks orbiting in the GPS constellation could be subjected to a similar analysis, although there we would have to accept the necessity of following a more complicated itinerary in tracking down the localization of indeterminacy, and the divergent realities that intersect there. The principal point, however, remains: clocks topologically localize metastability. The resolution of metastability is much more provisional in the case of the clock than in the case of the crystal. The clock *suspends* any final resolution of its metastability by localizing indetermination at key points. Through these key points, divergent realities interact with each other. In suspending resolution, it repeats it. The very stability of time as a recurring sequence rests on that localized metastability. Clocks are not alone in this; machines and technical ensembles effect localized suspension of indetermination.

MODULATION AS TEMPORAL MOULDING

Perhaps regarding the clock as a machine that resolves metastability by suspending it and maintaining it does not go far enough. The clock on this account is only a step away from the physical individuation of a crystal. It figures as a suspended or prolonged individuation which sustains a relation to both an energetic milieu (gravitation) and a social milieu ordered by repeatable sequences of marks. But if we regard clocks as carriers of clocktime technicity, then something more complicated is involved. A progressive genesis of clocktime occurs between the pendulum's oscillations and the oscillations of caesium atoms. Clocktime modulates itself through the genesis of different technical entities. We can view what takes place during that genesis by regarding clocktime as a way of staging *modulation* of oscillations. 'A modulator,' writes Simondon, 'is a continuous temporal mold. . . . [T]o modulate is to mold in a continuous and perpetually variable manner' (1964/1995, 45). Through the technicity of clocktime, modulation occurs at two levels.

First, Galileo's simple pendulum, whose naturally resonant oscillations gradually die away after it has been set in motion, undergoes quasi-continuous modulation in Huygens' clock. The temporal form of the clockwork is moulded by the oscillations of the pendulum. The temporal 'matter' of the pendulum requires the energy stored in the weight-driven clockwork. The clocktime produced by the pendulum clock in a sense has no fixed form or matter, since both the oscillations of the pendulum and the cyclical motions of clockwork are reciprocally interacting and adjusting each other. The pendulum modulates the clockwork, and the clockwork modulates the pendulum.

Second, the 'form' and 'matter' of clocktime undergo *continuous* development and variation through modulation. (These terms are taken up by Deleuze when he describes 'a very modern conception of the technological object' (1993, 19): such an object involves continuous development of form and continuous variation of matter.) The modulation is legible as we move from Huygen's clock to the GPS. The increasing rapidity of oscillation between Huygens' pendulum and the atomic clocks of GPS requires that what is modulated, the oscillating matter, changes. As Simondon says, 'the viscosity of the support is diminished as much as possible' (1964/1995, 45) when modulation occurs more rapidly. The almost sensible instants in which pendulum and escapement reciprocally modulate each other are replaced by imperceptibly rapid contacts between the oscillating fields of micro-wave radiation and electric potential fields of certain electrons belonging to caesium atoms. Although the reciprocity of modulations essential to clocktime remains operative, the localization of indetermination in the atomic clock-machine has now been displaced from oscillations coupled to the earth's gravitational field and redeployed in the less palpable, yet still localized, interactions of oscillating electromagnetic fields.

MULTIPLICATION AND INCORPORATION
OF DIVERGENCE

As a temporal moulding, clocktime is 'continuous and perpetually variable'. Clocktime could be seen as a kind of event whose 'harmonics' or 'sub-multiples' fold different layers or conjunctions of oscillation together. (Gilles Deleuze, as if describing the modulation that develops between the oscillations of a pendulum and the oscillating fields of the resonating caesium atom, speaks of an event as 'a vibration with an infinity of harmonics or submultiples, such as an audible wave, a luminous wave, or even an increasingly smaller part of space over the course of an increasingly shorter duration' (Deleuze, 1993, 77).) The question as to what triggers this multiplication remains.

Simondon writes that 'the individual technical object is not this or that thing, given here and now, but something in genesis' (Simondon, 1958/1989a, 20). Again, if we wish to diverge from the stereotypes surrounding the acceleration of clocktime, the technical specificity of a particular timing regime, the GPS system, should be understandable in terms of this genesis. As a positioning system based on atomic clocks, GPS confirms the inseparability of timing and spacing. More generally, it shows that what counts as time and place depends heavily on the kinds of technical mediations through which a given collective structures itself temporally and topologically. The 'trigger' for the multiplication of clocktime oscillations is neither extrinsic nor intrinsic to society. Rather, it resides in the recurrent play occasioned by non-coincidence between a collective's topological and temporal limits.

Because of the complexity of GPS as a technological system, I will focus on just one illustration of this point: the production of the GPS signal structure through modulation of the basic oscillation produced by atomic clocks. In one form or another, this signal, broadcast from each satellite in the GPS constellation, contains all the information needed for a local receiver to determine its map location. A block diagram showing the modulations that comprise the satellite signal structure conveys the multiplication and filtering of the oscillations derived from the atomic clock, now termed 'the frequency standard'. This simple shift in terminology shows that hearing or seeing a clock has become less important than the clock's often invisible and silent infrastructural role in directly regulating and synchronizing other technical elements, and indirectly coordinating disparate elements of a collective. In technical terms, the modulating fields of the encoding oscillators superimpose various streams of information on the basic clock signal by shifting the phase of the primary oscillator. Several different layers of modulation are superimposed within the coded and encrypted locational and timing signals transmitted by the GPS satel- lites. There is not just one zone of contact between the primary oscillator and what parallels the clockwork, the rest of the GPS system. Here the oscillations of the clock are modulated along divergent paths within the device. Signals coming from different sources continuously mould the primary oscillations (Kaplan, 1996, 241–3).[11] A diagram of signal structure indicates some of the 'harmon- ics' of the oscillations.

This layering of modulation corresponds to an incorporation of diverse realities not yet fully represented or accommodated within the time of Huygens' pendulum clock. His plan for a universal standard of length based on the oscillations of a pendulum relied on constant gravity everywhere on the earth's surface. The GPS constellation, orbiting the planet roughly every 11 hours 58 minutes on slightly elliptical orbits, cannot make that assumption. Its orbits and the propagation of its timing signals are perturbed by non-ideal and difficult to calculate factors such as the sun and moon's gravitational fields inducing tidal changes in the earth's gravitation field, solar winds impinging on the satellite, orbital deviations caused by the slow release of atmospheric gases from satellites made on earth, variations in ionospheric and tropospheric conditions, multipath distortion of signals and, until recently, *Selective Availability* (the deliberate manipulation of the satellite clocks and navigational data by the US Department of Defense to ensure that the highest levels of signal accuracy are only available to authorized users). The modulations present in the clock signal testify to this complicated intersection. The signal broadcast by a GPS satellite is not just a clock signal; it also describes the status of the clock itself, and includes current and predictive data about the satellite's location and the atmospheric conditions relevant to the propagation of the clock signal, as well as the encrypted signals required by the USA for national security purposes.

The simplest account of the complex structure of the GPS signal would be to say that GPS takes the variations of its milieu into account.[12] It would be possible to trace how the multi-layered modulation of the primary oscillations links geographical, meteorological, cosmological, military, economic and legal domains. The intersection of these domains with each other through various forms of feedback and reciprocal modulation in GPS constitutes another provisional structuring of the metastable technicity of clocktime. When compared to the pendulum clock, the increase in the rate of primary oscillations does not derive from a delocalizing or homogenizing dynamism intrinsic to modern technology. Rather, it stems from the articulation of different points of contact between human–non-human collectives and their associated milieus. 'Technicity', Simondon writes, 'super-saturates itself by incorporating anew the reality of the world to which it applies' (Simondon, 1958/1989a, 158). Out of this super-saturated state, particular structures precipitate. The rate of oscillation of contemporary clocktime indicates the absorption of a field of contingencies that were previously left open to chance, or that were previously subject to different kinds of treatment (for instance, social, political or cultural representation).

Super-saturated by that incorporation, the technicity of clocktime restructures the limits of the collectives it belongs to. It initiates points of contact between collectives and what lies outside them and, in doing so, establishes new limits, new pathways of action and affect within the collectives to which it belongs. The multiplying modulation of clocktime which underlies something like GPS (or the computer clocks whose oscillations are a fundamental component of contemporary digital technologies) moves the critical phases which in the pendulum clock mediated between the pendulum's movements and the clock's hands into a more complicated ensemble of mediations. The quantitative multiplication of oscillations signifies a topological complication in the structures of the collectives.

THE IMPROPRIETY OF CLOCKTIME

An account of the technicity of clocktime does not address the integration of clocks within past or contemporary cultures in specifically sociological, economic or semiotic terms. Clocks still carry meaning. For instance, Otto Mayr has shown how the clock as a metaphor of order, regularity, authority and the work of creation was particularly significant to natural and political thought for several centuries in modern Europe (Mayr, 1986). The 'Long Now' project shows that, as a metaphor of order, clocks still carry weight.

Rather, the ongoing genesis of clocktime can be read as provisional, localized resolutions of the metastability of a collective whose limits are not given in advance. Clocktime technicity, in the terms used here, refers to the way in which collectives absorb contingency within certain sequences of order and synchronization. The absorption remains incomplete because timing and spacing includes undifferentiated potentials whose ongoing individuation accounts for the interval between 1.0 and 9,192,631,770 oscillations/second. The major point of divergence from the evaluations of clocktime offered by Heidegger, Virilio, Elias and Latour rests on the notion that a collective cannot completely define its own limits because it is not completely in phase with itself. A kind of structural incompleteness or virtuality remains.

At the outset, I said that technicity refers to the historical mode of existence of technical mediations. This claim can now be refined a little. Technical mediations are not directly represented in cultural and language. They only figure obliquely in the processes whereby cultures or societies represent their own ongoing collective life to themselves. ('Technology', by contrast, figures hugely as an object of contemporary discourse.) Clearly, clocks as a technical mediation have a history. But clocktime technicity refers to something different. While still integral to the life of a collective, it is concerned with the ways in which certain collectives provisionally structure their belonging-together as an ensemble of living and non-living processes within temporal and topological limits that cannot in advance be fully lived or represented. Timing regimes – the pendulum clock, the atomic clock – represent changing distributions of those limits, and different ways of articulating divergent realities, living and non-living, with each other. Something like clocktime is necessary to the ongoing existence of our globalized collectives. But why say anything more about clocktime technicity? Why not treat the work of clocks as a strictly social coding or ordering of relations with a group, as Elias does? Only because the mutability and eventfulness of clocktime would then remain inexplicable. Conversely, why not regard clocktime as the symptom of a generalized and pervasive technologization, as Heidegger does? Because that would attribute an essential dynamism to an abstract entity, 'technology', to which societies would be passively subject. Clocktime neither stands apart from collectives nor is it completely coded within their social functions or purposes. Its mutability stems from the structural incompleteness of collectives themselves.

Attention to the technicity of clocktime potentially offers, by contrast with both the globalizing and sociocentric views of modern technologies, a more nuanced and historically deeper treatment of why our collectives are at once durable and unstable. Clocktime permeates temporality. It inflects the anticipation of a future and the appropriation of a past. Through the localization of specific kinds of indeterminacy, the ongoing modulation of matter and form, and the incorporation of divergent realities in timing ensembles, the technicity of clocktime figures as one way in which collectives provisionally stabilize their points of contact with what exceeds them, and also open themselves to ongoing differentiation.

NOTES

1. See Appendices III & IV of Howse (1980) for a brief technical description of the two clocks.

2. See Janicaud (1997, 189) for a discussion of the processes of concretization of an event.

3. The reasons for this are complicated, and will not be discussed here. The clock as an exemplary modern technological object cannot tell us anything essential about time, mainly because it is *technical*. It forms a component of the more general framework of technologies which materialize in response to the essence of modern technology, but which themselves are not essential (see Heidegger, 1977). This separation between the technical and the essence of technology will be discussed in Chapter 4. For a clear and accessible account of Heidegger on temporality, see Dastur (1998).

4. Historically, Elias writes: '[before Galileo] timing had been human-centred. Galileo's innovatory imagination led him to change the function of the ancient timing device [the clepsydra] by using it systematically as a gauge not for the flux of social but of natural events. In that way a new concept of "time", that of "physical time", began to branch off from the older, relatively more unitary human-centred concept. . . . The significance of this emergence of the concept of "physical time" from the matrix of "social time" can hardly be overrated' (1993, 115).

5. Time was, prior to Galileo, tightly woven together with law and especially with the state. The Roman calendar, the medieval *tempora* or hours, or the *computus* (the system of calculation used in the Middle Ages in the church to decide on what day important religious feasts would occur) had all been manifestations of the 'human-centred concept' of time. With Galileo, 'natural events' are brought within the domain of this time.

6. This brief assessment of Latour's understanding of the temporality of sociotechnical collectives risks missing its main objective: to present socio-technical temporality as a multiplicity of times derived from relations between different elements, rather than a laminar, irreversible flow dominated by accelerating technological progress. It would require a much more detailed engagement with the role that time plays in technical mediations to fully establish Latour's argument. Here I merely want to indicate that the status of instability or metastability still remains obscure and perhaps under-represented.

7. Today, as the Long Now Foundation considers designs for a clock that could keep time for 10,000 years, or roughly until the next ice age, with only Bronze Age maintenance technology, a mechanical oscillator such as a pendulum or spring, coupled to solar events, still figures as the regulating

mechanism of choice. The Long Now clock designer, Daniel Hillis, suggests: '[s]ince no single [timing] source does the job, use an unreliable timer to adjust an inaccurate timer, creating a phase locked loop. My current favorite combination is to use solar alignment to adjust a slow mechanical oscillator' (Hillis, 1999). This is the same technical project that Stengers and Gille describe as constitutive of autonomous time: the pendulum clock material-ized as a way to count and periodically restart mechanical oscillations. Furthermore, synchronizing solar and mechanical time is the explicit object of much of Huygens' own discourse on the globalization of clock times in *The Pendulum Clock* (Huygens, 1986).

8. For the method of synchronizing the clock with the earth's revolution see Huygens (1986, 23–5). For an account of how Huygens came to present a geometrical proof of the accuracy of the clock, see H. J. M. Bos's introduction to *The Pendulum Clock* (Huygens, 1986, 1–2).

9. Obviously, it also requires the wave theory of electromagnetic radiation, which again understands the phenomenon of light as an oscillation. In 1678, Huygens proposed the wave theory of light which allowed the propagation of light to be understood and rendered predictable.

10. In an unexpected and highly original move, Stengers and Gille analyse these moments of contact between the pendulum and the clockwork in strictly thermodynamic terms, as a dissipative system. They understand the motions of the pendulum and the moments of its contact with the escapement as a cycle which converts potential energy to kinetic energy, and energy to information. I will not reproduce their analysis here, but merely indicate that the outcome is a conception of the pendulum clock as a dissipative system. That is, they show that there is only autonomous clocktime, on the condition that energy constantly flows into and out of the system. The implication is that the ideal law of time as a linear succession of measurable durations rests on a complicated series of losses or dissipative flows of energy. These losses are not deviations from the ideal, but the indispensable condition of the functioning of the ideal as a norm (Stengers and Gille, 1997, 198).

11. The different data sources include the system that generates a unique code identifying the satellite that is broadcasting the signal, 'almanac' and 'ephemeris' data describing the current status of the satellite's orbit (uploaded from the ground control stations operated by the US Department of Defense), and 'errors' deliberately introduced into the signal by the Department of Defense to deny precise locational data to unauthorized users. The basic oscillations derived from the atomic clocks are divided, modulated and superimposed on each other at a number of different rates to produce the final signal structure broadcast by each satellite.

12. The structure of the signal, for instance, bears within it the history of late nineteenth-century contests between imperial nation-states over time standards. See Chapters 5–6 of Howse (1980) for the debates on the establishment of GMT, on which UTC (Universal Coordinated Time) is based.

CHAPTER 4

Infrastructure and individuation: speed and delay in Stelarc's Ping Body

An infrastructure occurs when the tension between local and global is resolved. That is, an infrastructure occurs when local practices are afforded by a larger-scale technology, which can then be used in a natural, ready-to-hand fashion. It becomes transparent as local variations are folded into organizational changes, and becomes an unambiguous home for somebody.

Star and Ruhleder, 1996

On UNIX computer systems, a modest diagnostic program called *ping* has long allowed users to determine how fast networks are handling information. It measures the propagation delay between the host computer where the program is running and any other accessible network address. To 'ping' a node on the networks is to measure how long it takes for a data packet to reach a particular address and return. On most days, the delay here on Sydney's academic networks is in the order of tens to hundreds of milliseconds for onshore sites, and up to ten times that for offshore sites. This delay fluctuates from second to second, depending on who or what else is present on the networks. On 10 April 1996, and on several later dates, the performance artist Stelarc, connected to several computers, modems, video monitors and speakers, oriented his body to an unpredictable series of delays extracted from *ping* data and transduced as electrical shocks. In *Ping Body*, Stelarc effectively unfolded himself as a living map of technical delays, so that the speed and volume of traffic on the Internet regulated the electrical stimulation applied to various points on his skin.[1]

In a specific, localized and dated event, *Ping Body* links together three concerns that have tentatively circled around each other in the previous chapters. Here, in a work staged through communication infrastructure, time, technology and living bodies come together. At this juncture, Stelarc's *Ping Body* performance allows an important question to be posed: who or what experiences technology? It gestures towards a kind of collective individuation which, following Simondon, I will term *transindividual.* While it may be asking too much of a single work, I suggest that *Ping Body* can help us see in more detail what the technicity of time and corporeality means. For any subject, 'mediation between perceptions and emotions is conditioned by the domain of the collective' (Simondon, 1989b, 122). As Muriel Combes writes, 'we see here that it is only in the unity of the collective – as a milieu in which perception and emotion can unite – that a subject can gather the two sides of its psychic activity [perception and affect] and coincide with itself in some way' (Combes, 1999, 59). The notion of the transindividual forms a crucial part of Simondon's alternative account of psychosocial experience because it links the emergence of collectives to something that is not fully experienced or perceived by individuals. A collective is a process of individuation emerging from beings who are not entirely themselves since they are transductive. Much of the second half of *L'individuation psychique et collective* (Psychic and collective individualism) (Simondon, 1989b) concentrates on the idea of the transindividual as a way of conceptualizing experience without either privileging a pregiven individual subject position or a structural totality at the level of society. The transindividual refers to a relation to others which is not determined by a constituted subject position, but by pre-individuated potentials only experienced as *affect*. Speaking transductively, the transindividual structures itself by resolving certain incompatibilities through the collective.

This understanding of affect and collectives could offer a different interpretation of technology in general which, when it is noticed at all, is inevitably felt to be exciting, boring, interesting, risky, dangerous, 'cool', or 'nerdy'. While Simondon does not explicitly link the transindividual and an experience of technology, an association between technical objects and transindividual resonates throughout his work.

The notion of the transindividual gathers together the abstract threads of temporality, corporeality and technicity and helps ensure that the collective historical existence of technology does not fall back on pregiven ideas of subjectivity, knowledge, power, nature or history. It moves the focus away from the split between devices and bodies toward a less visible but vital middle ground of material practices. Some steps in this direction have been indicated in the earlier chapters. The notions of *transduction*, *originary technicity*, *iterative materialization* and *information* (in Simondon's sense) all concentrate on a different understanding of the fabric of human collectives. The general idea of *transduction* suggests that a diversity of actors, interests, institutions and practices are articulated together through specific technologies. It implies that collectives individuate themselves technically. Bodies, artefacts and ensembles co-individuate at different levels. So, in Chapter 1, living human bodies were presented as always woven together with other bodies, non-human, living and non-living. What counts as the matter of a living body, its capacity to bear the imprint of social norms, was here seen to be contingent upon technical patterns of repetition involving non-living matter-taking-form. As we saw in Chapter 2, collectives vary in scale and topography. They articulate diverse realities with each other. Together, humans and non-humans negotiate what a collective can do and where its limits lie. A tool, when it is understood as a material practice, not only implies an intimate reconfiguration of a living body, but also entails a collective dynamic. Collectively, life suspends its limits, it staves off stasis not only through growth and reproduction, but through technical mediations. In Chapter 3, one important consequence of technicity was examined: here the ongoing mutability and eventfulness of technical mediations (particularly those associated with clocktime) was understood in terms of the articulation of collective limits. However, we still need to ask: does a transductive account of technical mediations support a different historical experience of technology? And, if so, who or what experiences it differently? The question of who or what experiences technology will be developed in the next two chapters.

The longitudinal or historical dimension of technical mediations has been extensively addressed in the work of Martin Heidegger. His affirmative engagement with technology is often overlooked in favour of a more familiar and limited 'judgement' on modern technology that is also scattered throughout his work. Although Heidegger's account could be shaped into a thoroughly negative evaluation of Stelarc and contemporary technology in general, a more interesting path might follow the linkages between temporality and technology in his work. Along with Simondon's notion of the transindividual, Heidegger's affirmation of technology in terms of temporality will guide my reading of *Ping Body*.

VIRILIO AND STELARC

When it comes to questions regarding the shape of a technological future, Stelarc's work elicits strongly symptomatic responses, ranging from enthusiasm to abhorrence. They tend to oscillate between technological euphoria and the fairly bleak assessments of recent technology associated with twentieth-century critical theory. Technologically literate and well-informed about current developments in technology, the writings of Paul Virilio are a good example of the latter. Several years before the *Ping Body* performance took place, Virilio described Stelarc's work and views on the relation between technology and living bodies. Stelarc's art is entangled in the deepening nexus of teletechnological and biotechnological processes. For Virilio, he is 'a willing victim, as so often the case with the servant corrupted by the master', of contemporary technology (Virilio, 1995a, 114). Stelarc's work exemplifies the processes of 'endo-colonization' that are shifting technical performances into ever more intensive engagements with living bodies. Virilio's broad thesis is that the conjunction of biotechnology and teletechnology reflects a critical phase or discontinuity in the increasing speed of technology (Virilio, 1993, 1998). At the core of the processes which accelerate information up to light speed, 'a final type of centrality, or more exactly, hypercentrality – that *of time*, of some "present" if not "real" time' (Virilio, 1995a, 106) prevails. The crisis or discontinuity centres on the loss of delay in communication. In these terms, when Stelarc stimulates his own nervous system with real-time signals arriving from around the networked globe, he testifies to and exacerbates the crisis: he contributes to a generalized loss of any sense of location or distinction between inside and outside, 'the essential notion of being and acting, here and now, losing all sense' (p. 107). The problem here is an absence of delay, experienced as light speed instantaneity. Speed immobilizes living bodies *and* over-stimulates those senses – vision in particular – directly exposed to teletechnologies.

Virilio writes that '[w]e are witnessing the beginnings of a type of *general arrival* in which everything arrives so quickly that departure becomes unnecessary' (1993, 8). In his view, speed induces a kind of stasis. Instantaneity, or general arrival, obviates real movement. During the nineteenth century, industrialization had mechanized labour, sub-stituting 'the technical for the muscular effort of the worker' (1995a, 119), but under contemporary teletechnological conditions, spatial exteriority and temporal futurity are under assault from instantaneity. He concludes: 'such an end implies forgetting spatial exteriority as much as temporal exteriority ('no future') and opting exclusively for the "present" instant, the real instant of instantaneous telecommunica-tions' (1997, 24–5).

Furthermore, in compensation for their tendency to render move-ment superfluous (beyond manipulating controls), teletechnologies resort to massive stimulation: 'it is now a matter of amplifying the subject's vitality through the impulses of information technologies' (1995a, 126). Stelarc's electrostimulation by delay times renders this compensating amplification explicit: 'The loss or, more precisely, decline of the *real space* of every expanse (physical or geographical) to the exclusive advantage of no-delay *real-time* teletechnology, inevitably leads to the *intraorganic intrusion of technology and its micromachines into the heart of the living*' (original author's italics) (p. 100).

Although it would be difficult to deny a deep and complicated connection between the geographical deterritorialization and techno-corporeal endo-colonization, this thesis is strongly magnetized by an absolute notion of speed as instantaneity and absence of delay. For Virilio, contemporary technologies have irreversibly crossed an abso-lute time barrier marked by the velocity of electromagnetic radiation. He suggests that 'today we are beginning to realize that systems of telecommunications do not merely confine *extension*, but that, in the transmission of messages and images, they also eradicate *duration* or delay' (Virilio, 1993, 3). Beyond the time barrier, the disorienting experience of speed leads inevitably to a rupture of the living interiority of the body. Speaking very broadly, implants, endocorporeal prostheses and perhaps biotechnology more generally flow for Virilio from the artificially induced immobility of living bodies. They compensate for a loss of time composed of rhythms of anticipation and delay. Virilio's response seeks to counter uncritical technocentrism. Faced with disorienting acceleration, a technocentric response embraces incessant upgrading of technical competence and equipment, and gleefully anticipates the fateful dissolution of existing human orientations and certainties into processes of biotechnical evolution. Stelarc himself often seems to echo these terms; for instance, he states that 'the body is obsolete' (Stelarc, 1997). Certain aspects of *Ping Body* and Stelarc's other works offer themselves to this kind of interpretation. However, both Virilio's critical stance, with its vision of the displacement of 'natural capacities for movement' by sedentary teleoperative inertia (Virilio, 1997, 16), and uncritical technocentrism uneasily share a presupposition: the thesis that a loss of distance and delay irreversibly leads to 'intraorganic intrusion' not only tends to presuppose an uncontaminated origin or natural interiority which can be intruded upon, it also attributes a hostile agency to technology in general. At the risk of caricaturing Virilio's work in particular, I think that his and the technocentric response risk a serious disavowal of the depth and complexity of human–technical involvements. Keith Ansell Pearson has argued in relation to technocentrism that the 'collapsing of bios and technos into each other is not only politically naive, producing a completely reified grand narrative of technology as the true agent and telos of natural and (in)human history, but also restricts technics to anthropos, binding history to anthropocentrism' (Ansell Pearson, 1997, 124). A similar kind of narrative often lies behind accounts which forecast the invasion of living bodies by machines. To speak of the posthuman body undergoing autonomous technological evolution amounts to the same thing as reading human subjects as the self-present agents of their own history.

SPEED AND DELAY

Is there another way of understanding *Ping Body*, which could show how historically sedimented and embodied temporality becomes inextricable from an ensemble of technical mediations? Despite certain problems of accessibility, translation and political risks, this is where Heidegger's work is significant and productive. It strongly links technology and temporality. No longer does technology just happen within history as just one event among many others. It has its own specificity as an event, and this specificity concerns its relation to time. Any experience of technology in terms of speed, disembodiment or disorienting collapse of space–time differences might be seen differently from the perspective of temporalization.

The problem of 'too much speed' is the central problem here. (We have already encountered it in Chapter 3 in the contrast between the hand-axe and the thermonuclear bomb.) Speed lies at the heart of any experience of technology. The mobility of digital information, so frequently invoked in recent times, undoubtedly promises (even if it does not actually provide) an experience of acceleration for those whose reading and writing habits took shape through books rather than screens. Perhaps more profoundly, biotechnology, with its heavy reliance on an informatic treatment of life, exacerbates biopolitical tensions (consumer resistance to genetically modified (GM) foods; debates over genetic research into cloning) to the extent that it alters collectively articulated rhythms of growth and reproduction. The disorientation associated with the speed of information and with the rhythms of technical change more generally will, or perhaps already has, become normal. A chance remains, however, as *Ping Body* shows, to experience this disorientation more thoughtfully. From this angle, we could ask: do technical mediations also provoke – at least at their inception – a deepened exposure to delay?

Speed is always relative to delay. Only a *change* in speed is ever felt as such. We have no experience of speed except as a *difference* of speeds. Derrida asks: 'are we having, today, *another*, a different experience of speed? Is our relation to motion and time qualitatively different? Or must we speak prudently of an extraordinary – although qualitatively homogeneous – acceleration of the same experience?'

(Derrida, 1984, 20). As he goes on to point out, this form of the question, by opposing the quantity and quality of speed, risks moving too quickly: it could be too fast, in that it rushes to see a break or difference, ignoring historical continuities or repetitions. But it also could be too slow in that it might miss seeing, hearing or feeling an unforeseen shockwave because it is concerned with uncovering continuities and repetitions. The general point is clear: there can be no sensation of speed without a difference in speed, without something moving at a different speed. Rather than opting for one side or other (radically different versus the same), it might be interesting to ask what the difficulty of deciding about the right speed means.

It is no paradox to say that speed actually is coupled to delay, to whatever remains incompletely synchronized in a given context. Delay is a consequence of the relativity of speeds. Within a certain modernist framework, delay, as lack of synchronization, implies a failing, fault or breakdown of some kind. In Virilio's terms, culture lags catastrophically behind technological processes. However, if different speeds were not co-present in a given situation, even if only as memories of a different speed, there could be no experience of acceleration or disorientation. To be located entirely in the present, or to have access to complete instantaneity, if it were possible, would be to feel neither delay nor speed. The coupling of speed and delay can be taken a step further. Understanding who or what experiences technical mediations as a change in speed can be framed as a question about what kind of transductive individuation is occurring. More specifically, it means asking whether everything changes at the same rate in any process of individuation. Rather than being simply destructive of the integrity of lived experience (as suggested by Virilio and others), the effects of speed may attest to a specific *dephasing* associated with a transductive individuation. From this perspective, the experience of speed could be more interesting.

The problem of thinking about how to read such constitutive delay is, I think, pervasively inscribed in *Ping Body*. The work was performed in order to be archived or 'uploaded'. This means that even its performance in 1996 was heavily marked by its reception at other times and places. The files can be downloaded from Stelarc's Web site or accessed on the *Metabody* CD-ROM.[2] *Ping Body* was and still can be viewed from four different, even divergent, angles. A schematic technical diagram of the apparatus which transduced Internet delays into electrical stimulation of Stelarc's body illustrates how the performance functioned technically. Computer graphics showing data gathered from the Internet, and quasi-geometrical computer-generated animations of dismembered body parts (also shown to the live audience) represent the relation between the signals filtered from Internet traffic and a living body. Video files show the artist in performance, capturing on-screen the gestures and movements that took place on a certain date, at a particular place. The video images were displayed on a large screen during the event. Finally, a set of text files document and date the performances.

The following discussion attempts to establish some kind of mapping between these four different diagrams or inscriptions of the event of *Ping Body*. It reads the diagrams as traces of heterogeneous realities articulated in the kinds of individuation staged through information technology. My principal concern is the problem of mapping *Ping Body* as an event which unfolds temporal structures and indeterminacies between diverse living, technical, mediatic and social bodies. The motivation is to at least complicate the anthropocentric and technocentric tendencies of Virilio's and Stelarc's own interpretations, and to suggest some more general points of orientation in relation to the speed of contemporary technology. As Bruno Latour puts it, 'Technological mechanisms are not anthropomorphs any more than humans are technomorphs. Humans and nonhumans take on form by redistributing the competences and performances of the multitude of actors that they hold on to and that hold on to them' (1996, 225). This alternate reading of *Ping Body* aims to show some elements of the process. It draws first of all on a set of Heideggerian motifs concerning time and technology.

TECHNICAL SUPPORT

There is every indication that Heidegger would have been no more interested in Stelarc's work than he was in radio, television, cybernetics, satellites, film or mechanized agriculture and concentration camps. He also places no great stock in technical or scientific knowledge in general, explicitly claiming that 'science does not think' (Heidegger, 1994). A diagram of the technical apparatus which staged the event would on many accounts, including Heidegger's and Virilio's, be the most inessential and indirect mode of access to *Ping Body*. It merely shows how desktop computers, modems, electrodes, video and sound equipment were hooked together to transduce *ping* results into electrical signals ranging between 0 and 60 volts on Stelarc's arms, legs and torso. While the apparatus attached to and surrounding Stelarc's body must have been clearly visible during the performance, the ways in which the components formed an ensemble distributed well beyond the site of the performance can be read more directly from the diagram. The diagram schematically figures an infrastructure – an information network – within which the work is located.

What relevance does a diagram of the apparatus have? It could be read in a number of different ways. If the technical apparatus is itself just the material support of the performance, a diagram of the support has an especially secondary status. With respect to broader questions about the event of modern technology, it would have little significance: 'so then, the essence of technological is also not at all technical', writes Heidegger on the first page of 'The Question Concerning Technology' (Heidegger, 1954/1977, 13), as if to confirm that the technical details of Stelarc's performance are irrelevant.[3] Why does Heidegger, who consistently says that technology constitutes the most difficult problem for thought today, and who insists that we must reconfigure our approach to technology to go deep underneath instrumental and anthropological concepts (technology as a means to an end, technology as a product of human action) apparently repudiate machines themselves? Why say that the essence of technology has nothing to do with the technical? Is he saying 'no' to technology itself as a growing cluster of interconnected machines simply because toying with the intricacies of complex modern machinery prevents the emergence of some broader perspective in which this complexity would make more sense? Heidegger directly denies this: 'To start with, it must be said that I am *not against* technology. I have never spoken *against* technology, nor against the so-called demon of technology' (Heidegger and Wisser, 1988, 25). Heidegger's oft-repeated statement that the essence of technology is nothing technical itself needs to be understood as something else, something ambivalent and troubled no doubt, but not as a simple and untenable negation of the artefactual complexities of technology. There is a definite blindness to or intolerance of the technical that imbues his questioning of technology. This concern has been raised by Derrida, among others.[4] As Bernard Stiegler has argued in relation to the opposition between the essence of technology and technologies, 'the whole question is whether such an evaluative distribution – according to which technology *is only on one side* (of an opposition) not being itself *constitutive* of individuation – is in fact still metaphysical' (Stiegler, 1993, 41).

Read generously, Heidegger's statement could mean that the technical support itself, which in *Ping Body* figures as a contingent set of technical objects for the reproduction (and archiving) of Internet delays and video images, obscures something crucial about technology: the radically historical and constitutive involvement of humans in the diverse unfolding of beings. Simondon makes the same point in a different way when he says, in relation to large-scale technical ensembles, we need to look beyond the machine itself to a milieu which always includes *life:* 'there is something of the living in a technical ensemble' (Simondon, 1989a, 125). The tissue of interconnections which the diagram shows includes connections terminating at a living body. That body, according to the diagram, exists at the same level as machines, components and networks. It is not above them, directing them or exercising power through them. *Ping Body* points to a life amidst the technical infrastructure, without which the ensemble could not be what it is. Stelarc does not direct this ensemble, nor is he simply enslaved by it. At most, we could say that he complicates it to the extent that he links different parts of the ensemble to each other. Without him, *Ping Body* as an ensemble falls apart.

If, as Heidegger argues in broad terms throughout much of his work, technology constitutes *the* predominant mode through which whatever exists presents itself today, then the critical question is how that predominance is constituted. The technical taken by itself occludes differences, but not because it enervates the human, as Virilio argues. Rather, it can make it difficult to see just how thoroughly enmeshed technical ensembles are with a particular kind of life, here represented by Stelarc's own living body. For Heidegger, technical entities are in a sense accidental, even inauthentic, in relation to any comprehension of our own involvement with them. In touching them, hearing them, in coming into palpable and visual contact with them, we lose touch with an exposure to them. The technical support is disorienting, not because it is seductive, banal or terrifying, but because it diverts attention away from an essential involvement with it.

The schematic diagram of *Ping Body* indexes something important concerning 'the technical'. The apparatus composed of computer, modems, video and sound equipment, 'stimbox' (the device that controls voltage applied to Sterlac) and electrode patches shows a complex interface between a globally extended technical infrastructure (that of the Internet, which lies off the diagram to the right), and a localized here and now, in which we locate Stelarc as an individual man. The diagram (see www.stelarc.va.com.au/pingbody/layout.html) indicates the necessary participation of living bodies in a technical ensemble: we see Stelarc outlined as a body, and we see video monitor screens, video recorder, amplifiers and speakers. Their inclusion implies an audience, even if this is only Stelarc himself, which *perceives*.

The fact that the technical diagram leaves out perception, gesture, location or experience, and replaces these with a simple outline of a body is significant, but it is not the only thing left out; so are multiple layers of technical structure and codes implied by the computers or video equipment. No technical diagram, no matter how much more detailed than this one (which is very schematic) can dispense with blank surfaces or open boxes in which the marks of other machines or living bodies are inscribed. Neither a living body, nor a perception, nor a machine can be fully figured here. The boxes and lines on the diagram suggest that a technical system is not a fully determined context.

It must preserve a degree of structural indeterminacy in order to work. It is a technical support only by being both stable and yet not fully determined. We can read the boxes and lines of various components as indicating relative degrees of closure and discreteness of the technical ensemble.

The diagram, in short, signifies that there is no such thing as the technical-in-itself, the purely technical support. The technical always exists in a relational context or milieu which enfolds certain specific degrees of indetermination along with determination. Every machine awaits specific responses in order to be the machine that it is. This incompletion structures the machine. Without a margin of indetermination, it would not be technical. A margin of indetermination or incompletion constitutes the technicity of a technical ensemble. The question is how these degrees of indeterminacy materialize through repetition in relation to adjacent bodies, machines and milieus. In the case of *Ping Body*, the schematic diagram suggests that the apparatus opens a number of different dimensions or trajectories at once: cameras, video monitors, projectors and recorders pertain to the visualization and recording of the gestures and actions of a body; modems and computers transduce local information into the globalized Internet; and electrodes and Stelarc's 'Third Arm' relate to the limits of gesture and touch. The technical ensemble affords resolution of these different trajectories. They circulate around the unfolding of *Ping Body*, configuring a mobile set of limits for whatever body eventuates. If the essence of technology is not technical, this is not because any exposure to the technical necessarily diverts us from the deep issue of how we are involved with technology. Rather the technical in itself does not exist as such.

THE VISIBILITY OF DIFFERENCES: ARCHIVE AND AUTO-CONDITIONING

The reason Heidegger insists on treating the *essence* of technology is to concentrate on the way in which technology remains what it is. This has a specifically temporal dimension involving 'differencing' or 'difference'. The historical mode of existence of technology is unthinkable without attention to these differences. For Heidegger, Virilio and many other recent responses to technology, contemporary technologies, especially the cybernetic technologies which they all regard as inaugurating an intensified phase of technical evolution, tend to render differences imperceptible. When Heidegger resisted the collapse of the essence of technology into the technical, he did so in the name of differences, and in the name of the diverse, manifold paths along which things, including bodies which are not just things, emerge. In very compressed terms, in *Identity and Difference* he writes: 'for us, the object *[Sache]* of thought is difference as difference' (Heidegger, 1957, 37). The whole question of technology is how to think about what technology *is* without thinking technologically; that is, without erasing differences. If he is not against technology, Heidegger must offer a way of thinking about the essence of technology in terms of differences.

The risk of indifferentiations specifically concerns what becomes technologically *visible*. In this case, as Virilio argues, Stelarc's living body risks appearing as merely technological. The audience of *Ping Body* could see a computer-generated animation of *Ping Body*. (The archived versions of the work found on Stelarc's Web site and the *Metabody* CD-ROM centre on this interface.) The animated display has two main zones. On the left, a simple outline of a human body is again recognizable. During runtime, various points on this body flash to show the activation of the electrodes on Stelarc's body by *ping* values. On the right, simple three-dimensional wire-mesh animations of a dismembered body move in accordance with the signals received back from the Internet. Along the bottom, a data window displays the Internet locations and time delays drawn from the *ping* data.

If we are to avoid seeing Stelarc as a willing victim of technological colonization, then these simple images of an outline of a body, and of awkwardly dislocated wire-frame limbs, must be legible as something connected to the essence of technology, *das Ge-Stell* (or Enframing). While insisting that the essence of contemporary technology is nothing technical, Heidegger called on a quasi-technological term to name it. In German, *das Gestell* means a stand, rack, frame, chassis, or, more obscurely, skeleton, as Heidegger explicitly points out (Heidegger, 1977/1954, 20/27). This is similar to the sense in which in English, one can speak of a skeletal structure as a 'frame'. With the possible exception of the skeleton, all these meanings refer to some aspect of an apparatus that *supports* or emplaces. (Samuel Weber suggests the translation 'emplacement': Weber, 1990.)

The word *das Gestell* is hyphenated by Heidegger as *das Ge-stell*. What would it mean to read *Ping Body* according to this articulated frame or rack, which names the thinkable essence of technology? For Heidegger the one thing that is to be thought through the essence of technology is the way in which technology is what it is. It orders everything it touches in historically specific ways. Heidegger says that the disjointed *das Ge-stell*, the essence of technology, is *die Weise des Entbergens*, the foremost way of revealing (Heidegger, 1954, 28). The manner in which almost everything – social relations, plant and animal life, machines, economic, ecological and cultural systems, etc. – displays itself today is specific to this historical conjuncture and its permeation by technical processes. To say that the essence of technology is a way of revealing or disclosing is to affirm that whatever comes into being today, whatever stabilizes or materializes as a contemporary product, body, construct, artefact or phantasm, emerges from processes synthesized through or as *das Ge-stell*.

There are two facets to this essential visibility. The first concerns how things persist. We might say they exist *archivally*. Technologically configured, things and events are retrievable. They exist as *Bestand* or standing-reserve according to Heidegger: 'everywhere, everything is ordered to stand by, to be immediately at hand, indeed to stand there just so it may be on call for a further ordering. Whatever is ordered about in this way has its own way of standing' (Heidegger, 1977/ 1954, 17/24). The event of *Ping Body* includes, as an integral component, its own retrieval. The apparatus which surrounds Stelarc codes, indexes and formats the event ready for retrieval and reproduction. The archival facets of *Ping Body* were already implied on the technical diagram in the form of various recording devices, and the figure that we are discussing is also drawn from an archival system (an Internet Web site or CD-ROM). This figure is extracted from a sequence of data produced during the performance, explicitly for the purposes of uploading and storage on the Internet, as well as for the 'live' audience. It resides on the network as a set of marks (information), ready to be delivered, reproduced or transported elsewhere. Accessing these resources is not difficult because they exist in order to be accessed and reproduced. Every level of their organization and inscription is premised on repeated retrieval and reproduction. All entities touched by contemporary technics are subject to this archival storage and access through codes and indexes. By implication, this process of enclosure and retrieval is not just what 'lifeless' systems of marks undergo. On the left-hand side of the animated display of *Ping Body*, a set of body parts, rendered as 3D volumes under two-dimensional projection, are configured as addressable resources in an archive. The living body too becomes a resource of elements to be retrieved.

The other facet of the modern technological regime of visibility concerns the *paths along* which things come to exist. Again, we might say, they emerge or appear through *self-regulation*. The archive unfolds itself or moves into visibility along paths that Heidegger describes in terms of a quasi-cybernetic logic of feedback cycles, activating certain sequences of events repeatedly: 'the revealing reveals to itself its own manifold interlocking paths, through regulating their course' (1977/ 1954, 16/24). This formulation shifts the emphasis away from any notion of a subject representing the world to itself towards an interminable self-regulation at work in modern technology. In contemporary terms, we might say that the mode of emergence of technical objects works 'autocatalytically' by constantly firing off new confirmations of the orderability of things (Kauffman, 1993).

In Heidegger's terms, *das Ge-stell* constantly animates itself. It displays quasi-cybernetic dynamics. It is a mode of emergence that feeds back on itself. The articulating or dismembering mark '—' typographically signals one possible source of movement when it distends the frame, *das Gestell*, into two limbs.[5] The instability that Heidegger inscribes in *das Ge-stell* can be understood as a topological transformation where in *folding*, something, the same thing, also *opens*. The opening, setting in the open or extension can be more or less understood in the terms we have just used to analyse the archive-simulation interaction in *Ping Body*. Contemporary technology sets things out in the open, on visible surfaces, pre-eminently today that of the digital screen. It multiplies inscriptions, as we have seen in earlier chapters, in order to compress other things. From Heidegger's standpoint, the levelling or homogenization associated with modern technology flows with the imperceptibility of this setting in the open.

What gathers or folds together is more difficult to parse. The whole question of technology for Heidegger is a question of whether and by what path we can think of opening out and folding together. The gathering-setting out [*das Versammelnde jenes Stellens*] of technological existence does not present itself as such, since it is what regulates disclosure, emergence or appearing – for us – in general. We cannot think it directly because of our involvement in it, because it affects who or what we are. As Heidegger says, the human 'stands within the essential sphere of *das Ge-stell*' (1977/1954, 24/27). Furthermore, '[the human] cannot take on a relation to it retrospectively [*nachträg-lich*]' (ibid.). This is a crucial point: we are not in delay when it comes to an essential involvement with technology. Lastly, we are, for Heidegger, not too late when it comes to thinking that involvement. 'Above all, the question of whether and how we specifically get involved with that in which *das Ge-stell* presences never comes too late' (1977/1954, 24/28). The relation to the manifold, to multiplicity, to folded differences, can never come too late. At least on this point Heidegger cannot be assimilated to the broad stream of thought which regards technology as alienated from culture.

GLOBALIZATION AND TEMPORALIZATION

When *Ping Body* runs, the display shows something of the interlacing of these two facets of the essence of technology. The flashing points that appear on the outlined body on one side of the display mark the return of ping messages from remote nodes of the Internet. The delays in response show that the *Bestand* does not attain the ideal immediacy that some of Heidegger's formulations, and most of Virilio's, imply. The flashing points on the outline of Stelarc's body signal perceptible delays. On the other side of the diagram, the partially dismembered and prosthetically stimulated limbs of Stelarc signal another problem: people are not autonomous in relation to the essence of technology.

According to Heidegger and many others, the technological evening-out of differences is becoming more or less normal. Norms are defined, which tend to globalize themselves and decontextualize differ-ences. As Geoffrey Bowker writes, 'one could perhaps describe globalization as an inevitably ever-incomplete attempt to impose a uniform representational time and space on a heterogeneous collection of lived spaces and histories' (Bowker, 1995, 55). For Heidegger, their normality threatens the singular and constitutive exposure of humans to an essential metastability concerning their role in the emergence of things. How does 'the human' [*der Mensch*] play a constitutive role in the domain of *das Ge-stell*? Heidegger's answer centres on *temporality*.

Technology can unfold as opening out into visibility only to the extent that we expose or suspend ourselves in relation to the gathering or folding it entails. We can be caught up in technically staged synchronization or immediacy only because we are not fully in the present, there all at once, or fully synchronized. The folding or gathering relies on a certain non-simultaneity or disjunction which we can call temporalization. There is not time itself in its flux on the one hand, and the modalities of consciousness through which it could be apprehended on the other, but a single process of temporalization, which has no existence apart from the folding together of humans and non-humans. Heidegger writes:

> The human stands so decisively in the wake of the provocation of *das Ge-stell* that he does not hear it as a claim, that he tires of seeing himself as the one spoken to and hence also fails in every way to hear how far he ek-sists from out of his essence in the region of an exhortation or address [im Bereich eines Zuspruches], and thus *can never* meet only himself. (1977/1954, 27/35)[6]

Heidegger is almost alone in touching on this point: if we did not exist (*ek-sist*, Heidegger writes) as a mode of exposure to the non-human, if our exposure to the metastable folding of being was not associated with temporal perturbations (we will soon return to these), nothing would happen, either technically or non-technically.

This affirmation of delay does not mean that people need to be out of date in their thinking about technology, or that culture should always follow in the wake of technological change. Rather, it means that there can be *no* technology as such without temporalization, deferral, anticipation and delay. The critical core of the 'emergency' (understood as both the perception of an *urgency* and something *emergent*) of contemporary technology for Heidegger consists of the possibility that delay would not 'arrive' or appear. In other words, catching up with contemporary technology – being thoroughly up to date would pose a threat to thought. In contrast to Virilio's emphasis on setting out the effects of a generalized arrival through technology, for Heidegger the non-arrival of any questioning concerning our specific involvement with technology is a high-level risk or a danger.

In itself, speed is not the danger. The danger is that the question of how technology implies delay will not be broached. How could the question not arrive? Only if it is not thought in terms of delay. The question of technology, of how to think about its essence, is never too late unless it arrives without delay.

Unfamiliar as it may be, and still tending towards a supra-technical identification of the human with thinking, this view of technology takes it seriously as a disjunctive and pluri-potent event. It gestures towards different ways of thinking about essence, more attuned to a radically contingent implication of bodies and time in technical mediations. Seen from this point of view, *Ping Body*'s fixation on delay takes on a very different complexion. The performance refers to a complicated historical actualization which occurs neither as a side-effect of human activity nor autonomously beyond the human. Just as there is no technical support in itself, there is no synchronization or simultaneity that does not also imply an exposure to something which resists immediate retrieval. A constitutional exposure to delay marks the real-time simulation, or any other kind of technological synchronization. The question now is how this exposure is articulated or *embodied*.

For Heidegger, language itself answers that question. As Bernard Stiegler observes, for Heidegger, speech or language bears within itself the originary temporality which opens a historical world for humans (Stiegler, 1993, 41). The concentration on language has profound consequences for the thinkability of technology as an event. The problem of thinking technology is basically, for Heidegger, a problem of the inability to hear at a distance, or to detect echoes. Modern technology muffles the originary temporality of speech. By rendering time calculable, it creates other storage possibilities which compete with and scramble the memory work of language. (Lyotard (1991) develops this point.) Technical action anaesthetizes sensitivity to the radical futurity embodied in linguistic heritage. Heidegger writes such difficult texts on technology because he attempts to retrieve the event of that exposure in thoroughly linguistic terms. Part of his project is to resist language becoming merely technical. This commits his work to a complicated web of linguistic resources overwhelmingly drawn from German and classical Greek texts. *If* we assume that time simply becomes calculable in modern technology, then something like a historically continuous linguistic heritage such as German poetry and philosophy would appear to be the only source of radical contingency left. Yet this antipathy to calculation remains far too simple and crude. Not only language opens an historical world for humans. The historical life of collectives embodies itself in technical mediations. The temporality of collectives is intimately linked to their technicity. While the stress that Heidegger places on the technological folding of time as temporalization remains crucial (because it does not allow contemporary technology to figure as a simple assault on the living), we need also to ask if it is possible to preserve this emphasis on temporality without grounding it solely within language?

FROM GESTURES TO COLLECTIVES: INFORMATION AND TEMPORALITY

In the video still of the performance, Stelarc stands frozen, webbed with cables and transducers, exposed to a spectrum of visual, tactile and auditory signals channelled by technical media interfaces (computer-controlled electrodes, video screens, computer-generated sound). Behind him, his own image steps back *mise en abyme* in a rear video projection. Stelarc's gestures are projected on-screen by part of the *Ping Body* apparatus, probably by the top video camera in this case. (In the animated sequence archived at Stelarc's Web site, the peculiar rhythms and abnormal contortions of Stelarc's gestures and movements make more sense of the dismembered computer animations of body parts.)

Most contemporary machines are arranged so that living bodies perform gestures in contact with certain surfaces of the machine (e.g. the keyboard, the mouse, the touchscreen, data-glove, spoken words). In information technologies, these gestures can be parsed as a sequence of commands that make sense within a given context of action. Stelarc's gestures are not functional, nor is their significance obvious. They do not point towards or do anything. *Ping Body* presents a living body whose quasi-involuntary movements flow from a sequence of highly serialized and networked movements of coded information (the data provided by *ping* running on the UNIX computer) as gestures. Stelarc perhaps feels the arrival of the *ping* data as a shock through the electrodes, but his gestures cannot be read as relating to some other purpose. This reversal of the usual relation between body and machine can be interpreted as the stimulation of an otherwise increasingly tranquilized living body. The consequence of such an interpretation is clearly negative. Quoting Virilio again:

> What lies ahead is a disturbance in the perception of what reality is; it is a shock, a mental concussion. . . . The specific negative aspect of these information superhighways is precisely this loss of orientation regarding alterity (the other), this disturbance in the relationship with the other and with the world. (Virilio, 1995a)

Heidegger also often speaks of the loss of distance and proximity. For both Heidegger and Virilio, Stelarc's gestures can only signal disorientation, specifically in relation to 'the other' and 'the world'. In Virilio's terms, the reversal staged by *Ping Body* would only confirm the endo-colonization of the living by the technological: Stelarc stimulates himself in order to compensate for the passivity and immobility induced by the arrival of data on the networks. His gestures touch nothing and no one. This interpretation ignores the fact that the *ping* data concern delays. Read in terms of delay, the reversal of the order of interfacing in *Ping Body* (the interface accesses Stelarc, rather than Stelarc accessing the interface) introduces a promising complication. *Ping Body* does enact instantaneity. Instead of gesture pertaining solely to what arrives instantly, to what is immediate or in contact, gesture opens on to a certain non-touching touch amidst the speed of telematic communication. *Ping Body* embodies speed and delay as a constantly changing disparity or divergence between touch and non-touch. This is not to say that *Ping Body* configures a posthuman body. In all their differences, living human and non-human bodies constitute a zone of interaction and relationality for *das Ge-stell*. They transduce between machines, which themselves participate in the individuation of the collective. In the process of transducing between machines, living bodies themselves are reconfigured (as we saw in Chapter 2). Simondon writes that 'the living does not come after, but during the non-living' (Simondon, 1995, 149). Insofar as they are *living*, bodies only suspend completion of the physical processes of individuation.

Because they suspend individuation or because they stage an ongoing transduction of divergent realities, topological and temporal complications emerge. 'Time,' Simondon says, 'springs from the preindividual just like the other dimensions through individuation effects itself' (Simondon, 1995, 32). Topologically, the transindividual is neither interior nor exterior to a body, but the continually folding and unfolding limit between inside and outside. Insofar as they live, bodies transduce. They are a transductive operation in progress.

The objectless flows of the standing-reserve in *Ping Body* – image, data, word, action – stream across the folds and surfaces of bodies, especially around hand, mouth, eye and ear. What kind of transduction might occur through the interactions staged there? Not only do such flows support anthropocentric illusions concerning technology as a human construct serving human purposes, they also reiterate forgetfulness of diversity and differences by plugging living bodies into technical apparatuses, and entail de-differentiation. Their intersection might be thought of in terms of how humans and technologies co-individuate. The disparities in timing in the gestures of *Ping Body* are essential to that historically contingent bodying-forth or bordering which co-defines what it is to have a body and to be situated in relation to others within a technosocial collective. Without that lack of synchronization, no body, living or non-living, could *be*. The disorienting event of *Ping Body* goes further than placing a living body under the control of telematically dispersed sources of information. It marks delay and disorientation as the possibility of there being an outside, of there being time or others. We certainly recognize the flows as information in the usual sense. They might also give rise to *information* in Simondon's sense: the signification which springs up when an individuation discovers the dimension according to which disparate realities can be articulated with each other.

There may be something too literal in *Ping Body*'s treatment of delay: electrical signals applied directly to the skin certainly signify an experience of delay without passing through language or signs. Stelarc's gestures stem from neuromuscular stimulation by data concerning the status of a communication infrastructure. Yet *Ping Body* reverses the usual occurrence of a gesture in relation to technical ensembles. There is here, as Virilio says, 'a loss of orientation regarding alterity', but this loss need not be negative. *Ping Body*'s gestures anticipate and respond to delays induced by the anonymous fluctuating volume of other gestures propagating on the networks. Despite their quasi-involuntary character, the gestures are triggered at intervals dictated by the speed at which information is moving on the Internet. That speed is significantly affected by the volume of information on the networks. In turn, this means that the rhythm of the gestures testifies to a shifting intersection of dispersed flows, criss-crossing between different elements of contemporary sociotechnical collectives. (Again, contrary to Virilio's interpretation, the disorienting effects of speed bear a relation to others within them.) It becomes impossible to decide whether Stelarc's gestures are his, or the involuntary products of *Ping Body* as a technical ensemble, itself interlaced with other places and people. The line between the two – the relation to others, as marked by the dependency of the *ping* data on what others are doing on the networks; the delay resulting from the current state of communication technology – shifts constantly and in ways that cannot be fully anticipated. We already include an anticipation of a technical delay within our gestures, and this inclusion means that a gesture can never be entirely human or completely meaningful. If that line between the technical and the other cannot be clearly drawn, does not the whole question of a loss of orientation due to excessive speed take on a different complexion?

TRANSINDIVIDUAL EMBODIMENT

The undecidable inclusion of the technical within the non-technical helps deal with a certain lingering problem associated with delay. Speed is experienced through inertia and delay. It is evidence of temporal complication and differences. In certain respects, *Ping Body* affirms the temporal complications of the collective. Rather than seeing humans as under assault by technology, the notion of transductive individuation lets us see how information implies an embodiment of a particular kind: the transindividual. The most difficult point in Heidegger's account of technology centres on the problem of how to think the *essence* of technology without lapsing back into notions of essence derived from metaphysics and an ontology of substance. The first section of the 'The Question Concerning Technology' had analysed why plausible and still common instrumental and anthropological definitions of technology (technology is a means to an end for humans, technology is something humans make or do) fail to respond to the event of contemporary technology. Those definitions neglected to address the essence of modern technology as a mode of emergence or eventuation. In fact, that neglect, which is no mere oversight, is the core complexity in thinking technology for Heidegger: 'technics is that which requires us to think what is usually understood by "essence" in another sense' (Heidegger, 1954/77, 30/34). The event of modern technology implies a different concept of essence: the essence of technics, *das Ge-stell*, is not a genus or general ideal, of which all actual technical mediations, including *Ping Body* would be species or instances. Rather, it pertains to a mode of eventuation.

The essential complication here is that everything that occurs technically tends to delay the appearance of its own mode of eventuation. Delay is intrinsic to the technological as a mode of eventuation. The essence of technics is, with regard to its thinkability, in delay. The autocatalysis and archiving characteristic of contemporary technical processes expose things to visibility at the expense of concealing how that mode of eventuation and persistence involves humans in something that they don't make or invent, i.e. something radically different yet constitutive for them. The human–non-human constellation of technology radiates both an energetic and a dislocating ordering in visibility, accessibility and synchronization while eclipsing a slender and yet essential involvement for humans. The constitutive delay affects any thinking of that involvement in the eventuation of things.

The problem is that the delays which *Ping Body* works with are 'technical' delays, while the constitutive delay that Heidegger refers to when he talks about responding to the event of modern technology is not technical. The millisecond delays in Internet response times seem trivial in comparison to the 'epochal' suspensions and delays Heidegger is interested in. Those delays are definitely located at the level of thought and history. They concern the interplay between anticipation and recollection which constitutes whatever is presented as real or significant within that collective. Recapitulated in terms of Foucault's 'critical question' (see the discussion which begins Chapter 1), delay is a condition of possibility of critical questioning: 'what is given to us as necessary' can appear as radically contingent only because we do not live solely in the present.

Delay *can* function merely as retardation. In Virilio's view, technology streams away from culture, and in doing so, tends to pre-empt or supplant it. For him, technological speed functions as a kind of absolute anticipation which blocks the arrival of anything other than that which it programmes. Worse, this speed has no direction or meaning in itself. If, however, technicity remains a condition of meaning or intelligibility, speed might not be fully intelligible. As Derrida writes in answer to a question posed by Bernard Stiegler, 'technology is not intelligible' (Derrida and Stiegler, 1996, 121), since anything that forms part of the conditions of meaning or intelligibility cannot be fully intelligible in itself. (Temporality and corporeality also have this 'property'.) Unintelligibility imposes a limit on thought, but it is not simply a nihilistic limit. The 'merely technical' form of delay found in *Ping Body* becomes important because without this technical delay there can be no sensation of speed. Speed can only be embodied through technical delays.

WHEN WAS *PING BODY* PERFORMED?

A fourth file relating to *Ping Body* records the dates on which the work was performed. *Ping Body* was performed in Sydney, for instance, on 10 April 1996. This date marks a day and a place, although perhaps not the day or place Heidegger had in mind for the eventalization of the essence of technology itself when he wrote that 'technology installs itself everywhere until one day, through everything technical, the essence of technology comes in the event [*Ereignis*] of truth' (1977/ 1954, 35/39). In this formulation, and many others like it, delay loses its constitutive charge if the 'one day' is understood as a future present, as a day on which a long-delayed recognition of human involvement with technology will finally arrive. The event of technology cannot be simply or directly *datable* for Heidegger, without asking under what circumstances something like a date becomes meaningful as an inscription of temporal processes.

The archived diagrams, computer images and video clips surround an event that occurred on a particular day. The event of *Ping Body* is actualized on that day. Perhaps that particular date, 10 April 1996, does not count any more than any other particular date, such as 31 December 1999. We could ask, however: does *Ping Body* say something about how an event takes place today, about *datability* under certain mediatic-informatic conditions? We have already seen the reversal of the usual order of interfacing and experience of speed that occurs through *Ping Body*. Stelarc's gestures are involuntarily inflected by measurements of the delay time of communication. Those delay times themselves relate to shifts in the volume of communication associated with a multiplicity of diverse realities: the alternation of day and night, the division of the world into a set of time-zones governed by datelines, the closing and opening times of geographically scattered financial markets, the fluctuating volumes of information triggered by highly contingent political, economic, cultural and natural events. All of these shifts take place within a milieu of signalling and transmission synchronized by dating and timing protocols. The event of *Ping Body* articulates those conditions on each other. Dating and timing protocols operate in each single facet of *Ping Body*, from the technical apparatus, through the Internet protocols which passed back the *ping* data, to the video loop which recorded, mixed and transmitted *Ping Body* on local and remote video screens as a 'live' event.

Who or what is *Ping Body*? The ephemeral monstrosity of *Ping Body* mutely gestures toward a mode of individuation intimately associated with technology as event. It lives on relays and delays within contemporary informatic collectives. Simondon writes that a 'collective only exists if an individuation institutes it' (Simondon, 1995, 165). The transindividual participates in the individuation of a collective. Clearly, *Ping Body* does not allow us to decide whether technological change is destructive decontextualization or a radically novel event. It perhaps allows us to say taht the way in which something comes about today cannot be fully rendered fully intelligible in terms which separate what bodies, machines, time or human and non-human others are.

At the simplest level, as an event, *Ping Body* articulates diverse technical, mediatic and corporeal-gestural realities with each other. Perhaps what it further brings into question is the way things happen today. It asks whether they occur as just such an articulation of diverse realities. Their emergent relation needs to be thought about. Broadly speaking, Virilio's thesis is that technological ensembles catastrophically collapse existing orderings of space and time. They colonize the living, and reduce or erase differences. It is difficult to see how such a position could register relations between different realities, let alone open up any kind of affirmative possibility. Insistence on an idealized simultaneity and the natural interiority of a living body can stand in the way of critical questioning. Although there are strong grounds to diverge from Heidegger (for instance, on the basis of his still too metaphysical repression of the technical), I think that the problem of thinking contemporary technological ensembles cannot easily circumvent his discussion of the essence of technology and the human in terms of temporality. Nearly every other critical account accepts speed as the 'substance' of modern technology, and then tries to find ways of slowing it down. Heidegger alone structures a response to technological speed in terms of constitutive delay, as a divergence at the core of historical processes. But is it possible to imagine an affirmative collective response to the event of modern technology emerging from a Germanic-Greek philosophico-poetic linguistic heritage?

Ping Body can be read as something more modest. It gestures towards a collective individuation associated with technology. We are collectively incorporating regimes of simulation, communication and retrieval ordered by *das Ge-stell*. Yet the technicity of these technical mediations cannot be rendered fully significant because they in part condition meaning. Only at temporal and corporeal limits can they be felt, and there perhaps only affectively. (In Chapter 5, one such contemporary limit-experience will be discussed.) For collectives centred on signifying processes, this can present compatibility problems. *Ping Body* is transindividual to the extent that it inhabits this incompatibility between affective and technical processes. It diagrams convoluted pathways or processes for the emergence of technological objects, and assays the tissue of their interconnection as living. Technical mediations may lack meaning yet still be mutable and eventful within the life of the collective.

NOTES

1. Video image uploads, text and technical details of performances in Australia, UK, the Netherlands and New Zealand, as well as most of Stelarc's writings, are archived at www.stelarc.va.com.au/, and also in Cook *et al.* (1997).

2. www.stelarc.va.com.au/

3. I will not attempt to summarize Heidegger's thinking on technology. Some acquaintance with Heidegger is unavoidable in the following discussion. An overview can be found in Dreyfus (1995).

4. Derrida formulated his uneasiness with this assertion of separation between technology and its essence as follows:

> It maintains the possibility of thought that questions, which is always thought of the essence, protected from any original and essential contamination by technology. The concern, then, was to analyze this desire for rigorous non-contamination and, from that, perhaps, to envisage the necessity, one could say the fatal necessity of a contamination – and the word was important to me – of a contact originarily impurifying thought or speech by technology. Contamination, then, of the thought of essence by technology, and so contamination by technology of the thinkable essence of technology. (Derrida, 1989, 10)

5. Distractingly, for our purposes, since it would mean broaching questions of language, translation and history, this articulation only makes obvious sense amidst the semantic folds of the German text. The usual English translation of *das Ge-stell* as 'enframing' omits the hyphen altogether. The broken syllables of *das Ge-Stell* reflect a deep articulation in the process whereby things come to stand as technically orderable entities. Here I will not trace Heidegger's semantic path, but will simply suggest that the articulation of *das Ge-stell* conjoins a concerted folding or gathering with an opening or setting out (Heidegger, 1977, 19–21; 1954, 23–5).

6. There are several problems in understanding what is said here. The text has been subjected to a compression that places serious obstacles in the way of its translation and comprehension. Whenever Heidegger sounds most inaccessible, whenever the archive of his thought is most resistant to expansion, a kind of condensation or dense patterning through interlaced philosophical, poetic and etymological registers is responsible. It is very difficult not to skip over the pleated surface of the text. While the textual compression *repeats* and *re-marks* that very manifolding of entities which Heidegger is interested in thinking, it also reinforces a commitment to spoken language as the primary medium of access to human history. This, for instance, is why the human ek-sists as the one *spoken to* or *addressed*. The unfamiliarity

of Heidegger's way of *speaking* about contemporary technology derives ultimately from his project of repeating and remarking *in language* the historical
complications and sedimentation of a human exposure to something other
than itself.

CHAPTER 5

Losing time at the PlayStation: real time and the 'whatever' body

> If anything, the modern collective is the one in which the relations of humans and nonhumans are so intimate, the transactions so many, the mediations so convoluted, that there is no plausible sense in which artifact, corporate body, and subject can be distinguished.
>
> Latour, 1999

Technological infrastructure can seem quite remote from the affective, perceptual and gestural modalities of embodied individuals. This chapter continues to emphasize the entwining of infrastructure and corporeality, but also focuses on the implications for individuals and collectives. The entwining stems from metastabilities that previous chapters have variously analysed at different levels in terms of materialization, information, timing and embodying-delay. If a transductive understanding of technical mediations is to be of value, it must also make some sense of how *individuals* fit with individuation. In other words, amidst all of the mediated and suspended indeterminacies open to an always-already technically mediated collective, we need to ask: how does an *individualized* individuation occur? In part, an answer to that question comes from the notion of the *transindividual* introduced in Chapter 4. It allowed Stelarc's *Ping Body* to be located on the boundary of infrastructure, collective and body, and sketched a general answer to the question of who or what experiences technology. It should, however, be possible to say more about how 'individualization' arises in a specific technically mediated context. This chapter will discuss how gesture and perception individuate a living body in the context of an online, real-time computer game.

Such games have a broad and diffuse significance within contemporary cultural contexts. They exemplify certain historic transformations broadly invoked by Giorgio Agamben:

> To appropriate the historic transformations of human nature that capitalism wants to limit to the spectacle, to link together image and body in a space where they can no longer be separated, and thus to forge the whatever body, whose *physis* is resemblance – this is the good that humanity must learn to wrest from commodities in their decline. (Agamben, 1993a, 50)

What might this 'good' mean in relation to the movements of digital information that support the rapid displacements of monetary value characteristic of globalization? Judged by the value standards of this good, the computer game, we might say, is a technical mediation in the service of the degenerate spectacle of information-images. It internalizes current and archaic phantasms of pleasure, violence and control through simple narratives, crude moralizing filters and forms of self-identification.

Even so, I want to argue that in the ways that they currently link images and bodies, computer games expose us to something more than a form of cultural damage, more than an infantile embarrassment or commodity in decline. For all the gender-specific, class-bound and racially stereotypical identifications they readily elicit (identifications that support and key into the ongoing commodification of computation and information), there are certain metastable configurations and rhythms associated with these toy artefacts. They may help think through the problem of individuation of a living entity in a technically mediated collective.

Returning to Agamben's citation, three strands are of interest:

(a) Agamben speaks of 'historic transformations of human nature'. We would have to ask what specific transformations are involved, and how is 'human nature' to be understood here. In Agamben's suggestion, these transformations concern the awkwardly termed 'whatever' body.

(b) Whatever these transformations are, 'capitalism wants to limit' them 'to the spectacle'. We might ask: how is this limitation – of the transformations – to the spectacle occurring, and why does capitalism limit those transformations? In the spectacle, especially given the ever-increasing role of real-time or 'streaming' media, the reduction of delay to the moment of transmission features pre-eminently. Information, as a technico-economic processing of indeterminacy, tends to reduce delay to the instant; it stages a coincidence between the occurrence, recording and reception of events.

(c) Finally, and more problematically, how is the linkage between image and body to be thought of within the context of these transformations? Why is that linkage so important for Agamben (or for us)? In what sense does it constitute a 'good', something that we might work for, in resisting an alienating separation between bodies and images? This separation is to be resisted precisely because it contributes to an uninhabitable commodification that eventually, as Agamben argues in later work (Agamben, 1998), targets life itself.

This third point will be a major focus here. Wresting the good from a commodity in decline, such as the computer game, would involve a form of *bricolage* that diverts it from its assigned function within the mediatized spectacle that Agamben, echoing Guy Debord's *The Society of the Spectacle* (1995), invokes here. In the case of the computer game, that function involves the direct interfacing of what is stereotypically regarded as a fairly unpromising kind of individual, the computer game-player, into circuits of information. The stereotype of computer games regards them as closed worlds, purged of differences, and mostly involving narratives that emphasize extermination of differences rather than affirmative engagement with them. Those circuits, however, also engender linkages between bodies and images which compose the mediatized whatever body.

Agamben's concept of the *whatever* subtly articulates a collective belonging-together, or being in common, which does not presume any substantial whole or unity. Agamben writes: 'decisive here is the idea of an *inessential* commonality, a solidarity that in no way concerns an essence' (p. 18). It designates a kind of being that emerges as a modal oscillation between proper and improper, in the same way that the wavering line of a hand-written signature runs between the proper ideality of identity and the improper contingencies of its particular inscription. For the most part, operating at a quasi-ontological level,

Agamben's account is not framed in terms of technically mediated linkages between images and bodies, between audiovisual perceptions and gestures. Yet, whenever he avers to the contemporary *ethos* of the spectacle, this linkage comes to the fore. In the context of computer games, we may need to renegotiate the passage between ontology and ethics as a linkage between bodies and images. The space in which I will attempt to locate these linkages is time: the time lost in play and conceived neither objectively nor as a predicate of human subjects or cultures.

The discussion that follows forms a 'subjective' counterpart to the 'objective' account of clocktime technicity in Chapter 3. What was at stake in that discussion was a way of understanding the dynamism of a collective without resort to an essence of technology or society. The tentative conclusions concerned the incompleteness of the collective in terms of metastable regimes of timing and spacing. The chapter argued that the effects of instability and dynamism associated with contemporary technology result from the incorporation within a collective of the fact of its own exteriority or incompleteness. The ongoing metastability of certain collectives reflects an incomplete structuring stemming from their interiorization of the exteriority of technical mediations as temporality. This interiorization can now be approached from a different angle: that of discovering in an embodied subject something that is neither individual nor common identity. In Simondon's terms, corporeality designates the pre-individuated reserves which constantly comply with and contest social norms.

PLAY, TIME AND HISTORICAL TRANSFORMATIONS OF HUMAN NATURE

The discussion is divided into three major sections, beginning with a theoretical contextualization of the links between play, time and historical transformations associated with the informatic spectacle. The following section sets out how a specific real-time computer game figures in that spectacle. A final section addresses the question of how certain complications or 'kinks' in linkages between images and bodies in computer games open on to the peculiarly collective singularity that Agamben terms 'the whatever'. In Simondon's terminology, this 'whatever' would be termed the 'transindividual', although he used the term without any direct reference to the role of technical mediations in collectives. The unstable mixture of discussing the specificities of a particular game *and* broader questions associated not only with the informatic spectacle but also with temporality and corporeality will, I hope, be justified by the appearance of some promising disjunctions in the timing of information. They suggest how an information consumer might, tentatively and inconclusively, succeed in belonging to impropriety as such.

THE TEMPORALITY OF PLAY: BETWEEN STRUCTURE AND EVENT

Let me sketch briefly both why I think that *play*, even in the form of computer games, might open a way into the problems of the historical transformations of human individuation; how, through play, we might both discern certain collective transformations associated with digital technologies and something important but elusive as to how one could belong to an improper yet singular collective. The argument will situate the computer game in relation to the time of the spectacle referred to by Agamben. Rather than treating the computer game as a semiotic process to be analysed according to its coding of narratives, or viewpoints, we might need to see it from the angle of temporality, and through temporality, historicity.

We could begin by regarding the computer game as a toy, something that 'children of all ages' play with. The sheer investment of time, money and energy that goes into computer games attests to how much time is given or lost to these toys by certain people, most prominently by boys and men, but clearly not just them.[1] A computer game is a form of play, and through it, the computer becomes a kind of toy. Games figure prominently in the marketing of personal computers for home use. Home computers are promoted on the basis of their capacity to stage the multimedia spectacle associated with contemporary games. No matter how commodified it is, or what economic value it has as merchandise, the computer game functions also as a form of play, or as a form of toy.

What is specifically at stake for contemporary social collectives in their status as forms of play, or as toys? There is a powerful double nexus between play and temporality, as Agamben himself has pointed out elsewhere:

(a) First, the history of games shows that play emerges from ritual, and ritual is deeply interwoven with time and history in the social formations it belongs to: sacred ritual maintains the calendar by marking thresholds between seasons in many societies, for instance. Compressing Agamben's argument a little, it could be said that rituals absorb events into synchronic structures, such as myths and calendars. Play and games, as desacralized ritual, enact rites as actions emptied of mythical content. Agamben writes: 'so in ball games we can discern the relics of the ritual representation of a myth in which the gods fought for possession of the sun' (Agamben, 1993b, 69). Whereas the narration of myths 'allows a sequence of events to be placed in a constant framework in which the beginning and the end of a story form a sort of rhythm or rhyme' (Lyotard, 1991, 67), play entails a loss of ordered time, or a breakdown of the time of the sacred. Play transforms structures into events.

(b) Second, at a less abstract level, play keeps on subsuming not only sacred behaviours, but anything that once belonged to human practice, whether it is practical, economic or military. Play even takes what still belongs to current practice, and miniaturizes it in the form of toys. Describing this aspect of play as transformation, Agamben writes: 'the toy is what belonged – *once, no longer* – to the realm of the sacred or the practical economic' (1993b, 71) – or to the military– industrial complex. Does this temporal dimension of the toy and play strongly emerge in the computer game? The computer game temporarily configures the computer as toy, and playing a computer game entails interacting with the computer as a toy. Support for this interpretation is not hard to find. Not only was the personal computer itself largely the invention of computer game-players in the late 1960s and early 1970s (in California),[2] contemporary computers are increasingly *miniature* in the sense that all toys are miniatures. Early *micro*computers transformed a current object – the large mainframe computers, belonging to governments, corporations, the military and universities – into a toy, a site of *bricolage* pervaded by televisual culture. (Hence

Bolter and Grusin (1999) can argue that there are no absolutely new media, only remediations of earlier media.) This 'toy' from California was sold back into those institutional domains a decade or two later in vast numbers, so that we humanities academics have computers on our desks today. The economic turned a toy to its own ends and this reappropriation increasingly structures the artefact at all levels of its fabrication and marketing. On the one hand, semiconductor manufacturers such as Intel design their products with the graphics-intensive requirements of computers games foremost in mind (e.g. by 'hard-wiring' certain mathematical transformations commonly used by programs that compute images representing the movement of objects in 3D space). On the other hand, in March 1999 under their 'FreePC' scheme, Compaq (the largest personal computer manufacturer) began to give computers to consumers in exchange for full information on consumption habits. These considerations, however, should not prevent us from asking whether the computer as toy, and the computer game as presentation of the computer as toy, can still share in the specific destructuring temporality of play. In terms of the example I will soon be discussing, this is to ask how *real time*, where the interval between the triggering of an event and its processing/reception falls beneath the threshold of sensible perception (i.e. faster than conscious thought), participates in this temporality. Real time can be understood as a touchstone of the ongoing historical transformations in human nature that Agamben refers to.

PLAY AS MATERIALIZED HISTORICITY

Agamben's analysis of the toy leads to a more general point as to the implication of play and time:

> What the toy preserves of its sacred or economic model, what survives of this after its dismemberment or miniaturization, is nothing other than the human temporality that was contained therein: its pure historical essence. The toy is a materialization of the historicity contained in objects, extracting it by means of a particular manipulation. (Agamben, 1993b, 71)

This is a claim strongly rooted in historical materialist accounts of subjectivity. If the toy somehow preserves the historicity of objects, if human temporality imbues it, and play extracts or exposes that temporality through manipulation, then there would be strong grounds to ask what kind of temporality, what materialization of historicity, can be discerned in computer games. In some way that needs clarification, computer play – and along with it, the computer insofar as it functions as a toy – would be a cipher of the temporality associated with our relation to information. It should enact something about what time is for us, in what sense we can be historical through and perhaps despite computation, as the goal of real-time transfer of information is relentlessly pursued in diverse spheres. When Agamben argues that historicity can be seen in play, he does not mean that play *represents* the way in which human temporality unfolds. Rather, he argues that the dynamics of that temporality unfold as play. In other words, play is not derivative, secondary to temporality. In particular, play would not be foreign to the temporalizing dynamics of humans in their exposure to history, but deeply embedded as a series of oscillations converting synchronic structures into diachronic events. Play inverts ritual: if ritual (or secular narratives of progress or emancipation which ground their legitimacy in the future rather than the past) absorbs events into synchronic structures, play in all its forms articulates and even dismembers synchronic structures into events. It doubles, replicates and displaces symbols, gestures and figures outside the codings which organize them within static structures.[3]

THE 'WHATEVER' BODY AND PLAY

From the standpoint of this perturbation of temporal structure, we can draw out some links between play and the 'whatever' body, links that pass through the 'historical transformations of human nature'. What Agamben understands by 'human nature' is adumbrated in *The Coming Community*. Devoid of any essence, historico-spiritual vocation or destiny, humans are and have to be their 'own existence as possibility or potentiality' (Agamben, 1993a, 43). If humans *had* to be something, ethics would be redundant or impossible. Because they don't have to be anything, only their potentiality to be (and their potentiality to *not* be) matters. The quasi-concept of 'whatever' figures this constitutive impropriety as a kind of constantly oscillating emergence, alternating between potential and act, between common nature and singularity.

> The passage from potentiality to act, from language to the word, from the common to the proper, comes about every time as a shuttling in both directions along a line of sparkling alternation on which common nature and singularity, potentiality and act change roles and interpenetrate. The being that is engendered on this line is whatever being, and the manner in which it passes from the common to the proper and from the proper to the common is called usage – or rather, ethos. (p. 20)

The kind of entity in question here is 'not a final determination of being, but an unravelling or an indetermination of its limits: a paradoxical *individuation by indetermination*' (p. 56). In consonance with much recent theory, this formulation acknowledges that singularity is not complete differentiation down to individual specificity. On the contrary, the singularity of 'whatever' is yet to be differentiated, yet to be analysed or filtered by dominant codings. Indeterminacy and deep contingency consist in a reserve of pre-individuation, or a constitutional openness that triggers becomings, invention and indeed play itself. To refer to 'historical transformations of human nature' would be to point to the process of enduring through this incompleteness, never being simultaneous or self-coincident, always being in some sense delayed or lacking synchronization with a milieu, and always running ahead of the present either through anticipation or perhaps openness to the unthought. History is the site of exposure *and* structuring articulation of this domain.

Through the notion of 'whatever' as the 'matheme of singularity' (a somewhat paradoxical term), I read a way of negotiating the double bind between improperness and properness, between a loss of particularity through an apparently irresistible global circulation of information and a fraught insistence on binding cultural specificity to historical and geographical context, that structures many accounts of the informatic spectacle. Jean-François Lyotard avers to this double bind, for instance, by observing that:

traditional culture thus remains profoundly marked by its local situation on the surface of the earth so that it cannot easily be transplanted or communicated. . . . The new technologies, on the other hand, in as much as they furnish cultural models which are not initially rooted in the local context but are immediately formed in view of the broadest diffusion across the surface of the globe, provide a remarkable means of overcoming the obstacle traditional culture opposes to the recording, transfer and communication of information. (Lyotard, 1991, 65)

Even if this diffusion is clearly incomplete and massively uneven (in terms of who has access), the double bind concerns what kind of culture could respond to this currently operating *model* of globalized diffusion.

The notion of the 'whatever' body offers a way to reconfigure this double bind in the context of the informatic spectacle. Its point of leverage is an altered understanding of the relation between singular individuation and the commonality or generality associated with information. This understanding allows focus to be maintained on the *ethos* of linkages between images and bodies, rather than trying to locate a new proper essence or identity which could elude the strictures of the double bind. On the line of 'sparkling alternation' between proper and common, particular and universal, those linkages compose an *ethos* which is neither proper nor improper, but 'a singularity without identity, a common and absolutely exposed singularity' (Agamben, 1993a, 65).

TRANSFORMATIONS LIMITED TO THE SPECTACLE?

How would such an historical transformation of collectives become legible? If play materializes the historicity incorporated in objects, then it should also be an exemplary site from which to draw some of the elements of the *ethos* of the 'whatever' body. Play sometimes occurs as just that kind of modal oscillation between structure and event, universal and particular, which the 'whatever' entails.

In very broad terms, collectives order time according to the differential relations they establish between event and structure. At one extreme, a collective of play would be an impossible society of the pure event; all structures would be translated into events, into a loss of ordered time. Chronological and calendrical time would collapse. At the other extreme would lie a society of the eternal present, knowing no difference between past and present. In any given society, specific mechanisms of ritual and play incessantly transform structures and events into each other. Given the impossibility of existing at either extreme (i.e. a society without structure, or a society without event), historic transformations in human nature have to be seen in these terms as modifications in the rhythms, rates and repetition of movements between structure and event, and event and structure.

Where would the computer game lie between these two extremes? In what way does it transform structures into events? Turning to the specific example I promised at the outset, that of a recent real-time animation game, how does it artifice the margin between structure and event? How does it preserve, as a toy, the temporality captured from the sacred/economic model it dismembers? Finally, in the context of the informatic spectacle, how does it limit the historical transformations in which we are enmeshed?

REAL-TIME SYNTHESIS AND CODING

A game called Avara exemplifies the currently dominant genre of computer games: real-time animated combat, a style of game that excites strongly gender-specific interest and drives the frenetic development of faster graphics-intensive computation (see www. ambrosiasw.com/games/avara). As a real-time, networked action game, it tightly binds together considerations of time, speed, bodies and information. For players, the main game scene displays an image of self, an other, an alien (by definition, a hostile other), and a 'closed-world' setting (Edwards, 1996, 3). As a game played over networks, not only are there computer-controlled opponents (the aliens), but also other human-controlled players in the image. Actor and spectator are interchangeable in this spectacle whose main theme is surveillance and security. In Avara, each stage or level of the game introduces a new scene in which hidden thresholds or traps complicate the surveillance activity, allowing other players to see without being seen. As

Bolter and Grusin have recently argued, in certain ways this type of game shows the influence of other media such as television, video, cinema and the novel:

> Like television, these games function in real time: either the player tries explicitly to 'beat the clock' or faces some other limitations, such as the amount of ammunition, which defines the rushed pace of the game. Finally, like television, these games are about monitoring the world. . . . [P]layers of action-style games are called on to conduct an ongoing surveillance. (Bolter and Grusin, 1999, 63)

In Avara, furthermore, a player usually meets other players on the networks. The circuit of hand-keys-computer-screen-eyes is not only mediated by the labyrinthine paths of the software, but also by the protocols and topology of digital networks. The time and space of the game is enmeshed with the globally extended but unevenly distributed passage of information. Whatever images are generated in the course of this real-time animated game, they are accessible to others, perhaps from different standpoints. The play is not only located in the manipulation of certain image-objects through keyboard controls, but in an interaction with other players, whose gestures and bodies are figured as objects on the screen. By contrast with other well-known games like Myst, which are based on a series of linguistic puzzles entwined with photoreal stills of mythical scenes, game play in Avara centres around *real-time animation*. Whatever the game may embody in relation to temporality passes through the synthesis of real-time animated images by the game's 'graphics engine'. The software acts as a machine to constantly redraw a geometrically defined set of entities on the screen within certain temporal or, more precisely, chronometric parameters. That is, the images have to change fast enough to fall beneath the threshold of players' conscious perceptions. The software of the game seeks to keep its rate of redrawing above the minimum rate of around 25 frames per second, the same constraint that both cinema and television must meet in their own ways.

The way in which the game synthesizes images within these temporal parameters is by coding the scene heavily around a mobile viewpoint occupied by the player. These games are called '3D-immersive' because the polygons mapped on to the 2D screen are generated within a three-dimensional Euclidean geometry, and then mapped back on to the screen according to axioms of linear perspective developed in Renaissance art. As opposed to earlier games such as Space Invaders or Pacman where the scene had no depth, Avara and its contemporaries such as Doom, Quake, Marathon and so on increase players' sense of mobility and speed by mapping it according to the rules of linear perspective.

Geometric perspective is crucial to the constitution of an almost arbitrary vantage point from which the space beyond the screen can be surveyed or controlled. Perspective allows much more complex orderings of objects, more diverse spaces, and more complicated and dynamic trajectories through them. Moreover, as Bolter and Grusin observe, such games implement linear perspective so thoroughly (through the mathematics of linear algebra and projective geometry) that there is no room in the game scene for 'the distortions or deliberate manipulation of conventions that occur in Renaissance painting' (Bolter and Grusin, 1999, 26). Indeed, any such manipulation is made difficult by the fact the graphics engine can only maintain the frame rate by computing all visible entities as collections of straight-sided polygons. What appears to us as a player figures on the screen as the continual rotation and translation of a few dozen polygons held in proximity. The game scene, including everything that is either stationary or moving, is uniformly composed of polygons. There are no curves here, no variations in texture. (Avara shows its age here, because most current action games rely on 'texture-mapping' to furnish the game scene with variations in texture.) The peculiarly synthetic appearance of Avara's world, and most animated computer games, stems from this reliance on polygons as the low-level building blocks of images.

Polygons are particularly appropriate for the construction of per-spectival views because their linear geometry can be easily transformed to represent the diminution in size associated with depth or distance into the scene. Furthermore, their well-defined edges permit the figures and shapes that move on the screen to be illuminated direction-ally. They have bright and shadowed sides. This play of light and shade is no reflection of light touching any body in the world; light too is coded into the polygonal geometry of the scene. The play of light and shadow gives depth and direction to a visual setting which would otherwise appear very disorienting.

What does this brief and perhaps familiar survey of the technical coding of the computer game scene indicate? How does this system of coded marks which composes this visual space (and the audible effects which I'm leaving aside) consisting of polygons, perspectival depth and directional light, extract any differential margin between structure and event in contemporary social collectives? In what sense is this kind of play temporalizing?

In general terms, we could say that the kinds of technical considerations just referred to increase the capacity of the system of marks known as digital computation to control and regulate events, to insert them in a frame. As a technoscientific toy, does Avara not conform to Lyotard's description of how information is processed based on a model of exchange dedicated to neutralizing events? That is, as Lyotard observes:

> Complete information means neutralizing more events. . . . [I]f one wants to control a process, the best way of doing so is to subordinate the present to what is (still) called the 'future', since in these conditions the 'future' will be completely predetermined and the present will cease opening onto an uncertain and contingent 'afterwards.' (Lyotard, 1991, 65)

In terms of the desynchronizing indeterminacies of play in general, as Agamben describes it, events at the interface between player and machine are *more* synchronically determined within games programs, rather than less synchronic. That is, any event diachronically triggered by a player can be processed and conveyed to self or another player solely within the 3D perspectival space composed of coloured polygons. The coding of a scene in that geometry and as a collection of polygons forms a highly organized structure that can absorb almost instantaneously almost any event that propagates into it. There is no visible motion within the space of the screen that has not undergone processing in terms of this computational geometry and the linear translations coded through it. The very experience players have of being in control, of being able to play the game, depends on synchronic structures that anticipate all possible moves in advance, structures that are dynamically repopulated with new data by each move in the game, but which themselves are not directly exposed to play or random *bricolage*.

Rather than converting structures into events, the real-time animated computer game seems to assimilate events to pre-existing structures, to select amongst the possible events only those that can be processed in terms of translations of polygons around the screen. Rather than real-time play triggering events, the very systems on which it relies seem to contain events within a strictly controlled permutation of marks. There would be good grounds to argue that there is no play here, or at any other 'playstation'. The structuring of information as value governs the field of play in a way that Lyotard again describes clearly: 'According to this way of treating time, success depends on the informational pro-cess, which consists in making sure that, at time t', nothing more can happen other than the occurrence programmed at time t' (1991, 66). From this perspective, there is no future in computer games. To play a computer game represents a loss of time, since nothing happens, *except what was programmed*.

INFORMATION AND THE VALUE OF PLAY

Furthermore, not only is this movement of polygons modelled on exchange, it is also a race to extract value from information. A rough equation between moving more polygons per second on screen and the game machine as a commodity form governs profits in the games and expanding online entertainment markets. More polygons per second means more visual detail, and a greater sensation of speed. Games such as Quake and Doom, or machines such as the Sony PlayStation represented quantum leaps in capacity to compute polygons per second. When consumer advertising talks of excitement, speed and action, it refers almost solely to the capacity of a particular game or machine to mobilize polygons on-screen. As a commodity form, the polygonal scenes of real-time games instance a more general rule which holds that the value of information equates to the time of its circulation. Once information is everywhere, fully diffused in the networks, it loses value. For that reason, the informatic spectacle must constantly inscribe new relative delays in the circuits of information. In relation to the telecommunication of finance, Gayatri Spivak has commented that '[e]ven as circulation time attains the apparent instantaneity of thought (and more), the continuity of production ensured by that attainment of apparent coincidence must be broken up by capital: its means of doing so is to keep the labor reserves in the comprador countries outside of this instantaneity' (Spivak, 1996, 123). While the money-form of value might attain real-time instantaneity in certain privileged and highly invested domains (such as the foreign exchange markets), there must still be a speed differential to be exploited. To have something faster, sooner, now rather than later, is what defines the value of information as a commodity. Hence, the only thing that is of value is a *relative* reduction in transmission time of information. Insisting on value, the very principle (and principal) of capital resides in different delays. Although the value of information rests in its speed, this speed only makes sense as a differential. There must be differences in speed for information to have any value.

By promising instantaneity between an event and its reception, real time seeks to eradicate delay. Yet at the same time, there must be delays somewhere, otherwise capital would not deploy itself in real time as information or live spectacle. There must be different speeds of access to information or different rates of movement of information if capital is to market the spectacle to consumers as a form of merchandise. The only solution to this necessity is to continue to speed up, to stage differences in speed by circulating information to some places faster, thereby reinscribing relative delays in movements of information as the source of value.

Indeed, the most significant transformations that capitalism wants to limit to the spectacle would be precisely those unstable transformations associated with real-time movements of information. Guy Debord, in *The Society of the Spectacle*, wrote that the 'time of the spectacle' is the 'time appropriate to the consumption of images, and, . . . the image of the consumption of time' (Debord, 1995, 112). We would need to ask in the context of the computer game how 'a time appropriate to the consumption of images' is figured, and how the game is an image of the consumption of time. For the sake of calculable delays, the diffusion of information must be manageable as risk rather than as indeterminate or incalculable delay. To resist that management, the question would be: what kind of incalculable or indeterminable delay can be associated today with real-time movements of information?

DISJUNCTIONS IN COMPUTATIONAL SYNTHESES: LINKAGES BETWEEN BODY AND IMAGE

Where, *contra* Lyotard, such coding and determination of a flux of images might also increase the reserves of indeterminacy is precisely in the domain of the linkages between bodies and images. The question remains: can a more complex and deeply contingent temporality inhabit the linkages of body and informatic images? How would this temporality work against the linear coding of images and the extraction of informatic surplus value through relative differences in speed? Given these strictures, how could we access any kind of historic transformation? Does not the very form of the technical coding involved in these games (which is of course patented) through its performativity neutralize any possibility of thinking through computer game play in terms of 'historic transformations' except as a commodified spectacle? Do not the patented processes subsume play within a generalized form of spectacle which relentlessly and incessantly re-enacts desacralized structures captivated by economic exchange value more than anything else?

This would be to ask in what sense, if any, these linkages can be thought differently as neither proper nor improper, but as the 'whatever' body. Such a body is neither ineffably individual nor intelligibly universal. The 'whatever' body would have to be, as Agamben points out, in effect an inhabitant of limbo, and neutral with respect to either salvation or damnation, *impassible* with respect to divine justice (or any of its secular heirs). The 'good' associated with the 'whatever' body would not be somewhere else, apparent from the linkages between images and bodies. Rather, 'it is simply the point at which they grasp the taking-place proper to them, at which they touch their own non-transcendent matter' (Agamben, 1993a, 15).

Let us try again to access this 'taking-place' in terms of the temporal instabilities of real-time play. The real-time synthesis of images entails different forms of anticipation and delay than those found at work in cinema and television, or in any other media that Avara re-mediates. There is nothing new in the requirement that animated computer games redraw screen images at least twenty times per second. Film and television already observe that constraint. Cinema and television rely on a coincidence between the succession of images synthesized by the apparatus and the flux of perceptions experienced by the spectator. However, real-time computation seeks to interleave gesture within the circuit. Without wanting to naturalize touch as the last vestige of proper corporeal presence amidst technological dislocation, as they come into contact with images, gesture and touch do make explicit some important disjunctions in the computational synthesis of spaces and times. From the perspective of controlling or stocking events, the presence of gesture poses problems that can be seen from both technical and phenomenological perspectives.

(a) *The technical-corporeal problem.* The effect of immediacy generated by the televisual apparatus had, until recently, no real possibility of immediately reconfiguring events on-screen according to viewer response. By contrast, the real-time game must devise some way of remaining open to players' gestures. In the real-time computer game, a transduction occurs between the indeterminate sequence of gestures players might be induced to send across the interface and the supposedly stable constraint that visually coherent (i.e. not fragmented or incomplete) images are remapped at least twenty times per second. The computer game software configures the computer as a transducer that articulates gestures and images. This capacity of the machine to be repeatedly determined or 'informed' at any time is called 'interactivity'. Certain crucial portions of the program must remain sufficiently indetermined to accept a limited range of contingent events coming from the program interfaces. The technical problem is to ensure that the transduction of indetermination into the determined form of images occurs within an interval more or less beneath the threshold of conscious perception.

During each moment of play, gestures mingle with images. Every gesture in a game like Avara either changes a player's location or marks his attempts to 'touch' another player.[4] But the intermingling of bodies and images does not take place on-screen itself or, at least, no more than the writing of a book takes place as its pages are printed. Like reading and writing, it takes place in a space where the lines between the particularity of a gesture and the coding of a mark are difficult to trace. The singularity of this event can be specifically located in the case of Avara in its binary space partitioning trees, the patented data structures which transduce touch into geometrical form, and where the set of polygons composing the visual space of the game is dynamically structured to permit interaction between player viewpoints and player gestures. If to have a body is to be open to technical mediations, then having an Avara body involves the particular pathways and rhythms of action and perception transduced through these structures.

In constantly melding gestures and images, Avara makes use of a binary space partitioning tree (BSP tree) to bring together gesture and image. A quasi-technical definition of the BSP tree reads:

> A Binary Space Partitioning Tree (or BSP Tree) is a data structure that is used to organize objects within a space. . . . A BSP tree is a recursive sub-division of space that treats each line segment (or polygon, in 3D) as a cutting plane which is used to categorize all remaining objects in the space as either being in 'front' or in 'back' of that plane. (Fuchs et al., 1980)[5]

The important feature of this somewhat technical definition is that every object in a space structured by BSP treatment is categorized according to its position in relation to a set of arbitrarily selected partitions or thresholds. Gesture is not directly processed within an homogeneous geometric continuum, but marked in relation to thresholds or partitions that classify every other entity within the scene as either 'in front' or 'behind'. Every movement in Avara, and the many other contemporary games that use the same techniques, takes place through traversal of a binary tree-structure of virtual viewpoints rather than through a continuous translation between points in a geometrical continuum.

The mechanism of the BSP structures the game space in the interests of visually mapping an indeterminate set of movements and intersecting trajectories, of tracking dispersions and fragmentations, and of permitting shifts between different viewpoints to be represented without discontinuity or comprehensive recalculation of the relations between entities in the scene. A player's viewpoint at any given time is defined by sorting all the polygons in the scene according to whether they stand in front of or behind that viewpoint. That these discontinuous traversals of the nodes of a data structure can be mapped back on to a continuous linear and perspectival geometry where objects seem to be separated or colliding, visible or occluded, should not prevent us from remarking the specificities of the locale where gesture and image are linked. In these terms, 'touching' means simply 'no longer recursively classifiable' as either 'in front of' or 'behind'. To see the game space as Cartesian, for instance, would be to go straight past the taking-place of gesture in Avara.

The BSP tree algorithms treat gestures as an occasion to re-sort relations between objects into the categories of 'in front of' or 'behind', touching or not touching, intersecting or not intersecting between objects. The separation between body and image comes into question here. Agamben observes that 'never has the human body – above all the female body – been so massively manipulated as today . . . And yet the process of technologization, instead of materially investing the body, was aimed at the construction of a separate sphere that had practically no point of contact with it: what was technologized was not the body but its image' (Agamben, 1993a, 49–50). If, in this context, bodies and images can become the 'whatever' body, something whose singularity is neither the ineffable particularity of a single body nor the universality of a technologized image, then it should be possible to regard the 'trick' that BSP trees perform with gestures and images as possessing a positive, albeit disjunctive, aspect. The BSP trees inextricably code gestures and polygons together. Within the context of these data structures, bodies and images are enmeshed. No doubt this inseparability stems, as mentioned above, from the temporal demands of real-time processing. The BSP trees render tractable a computational task that would otherwise require much more powerful computers; yet they also provide a glimpse of an ethos akin to the 'whatever', where images and bodies can no longer be separated. An example of the curiously decoupled intersection of touch and sight in this space is played out in Avara when, as sometimes happens, a player occupies the one place of total invisibility and non-touching contact in the game scene: the 'head' of another player.

(b) *The phenomenological problem.* The linkage between bodies and images that Agamben affirms as heterogeneous to the spectacle involves a *physis*, a manner of folding and unfolding of a body in relation to touch, sound and vision, not just image and sound images. In strict terms, to speak of the 'whatever' body as *physis* implies temporally complex relations between movement and rest, between change and stability within that body.[6] The complexities in the context of real-time computation occur at two levels: not only does a gesture somehow touch the image amidst labyrinthine passage of data, it also encounters the touch or the mark of others within the image. Gestures, as they pass into play, are the locus of an indeterminacy or modal oscillation in the taking-place of the 'whatever' body that exceeds the coding capacity of Avara as a system of marks. The ways in which gestures address an other, even a non-human other, constitute a form of openness or an element of play that is not fully processed by the synchronizing codes of the game. The event wherein my touch marks the image of another, and that others mark mine is another possible indetermination or site of ongoing individuation in an otherwise highly regulated or coded context. Clearly, the marking of gesture in the game, such that now the timing of touch (experienced as tactile contact with the keys and in proprioceptive perception of the hands) must encounter a flux of images, does not imply any necessary increase in indeterminacy. Yet how this indeterminacy can be related to the singularity of the 'whatever' body remains unclear. Ideally, in real time, things are supposed to happen to give the effect of 'liveness', of immediacy, of no delay between the occurrence of the event, its transmission and its reception. The occurrence of the gesture and the visual perception of it should be synchronized if the game is in real time. There *should* be no loss of time between touch, sight and sound, between the advent of an event and its reception.

However, despite the closed settings and limited focus, despite all the effort put into coding gestures, localities and images in advance, games like Avara are exposed to the fluctuating delays endemic to the circulation of information more generally. Sometimes Avara brings up the on-screen message 'Reality fragmentation detected' to warn players that the cross-mapping of image and gesture in the topology of the BSP is no longer in agreement between different nodes on the network. Occasionally, the timing of touch unexpectedly does not coincide with the flux of images. This discrepancy in timing can be regarded as an artefact of the 'incomplete' development of network technology. Although 'the time of spectacle' promises instantaneous delivery of information, that promise itself has not yet been and possibly will not ever be fully delivered, for structural reasons already mentioned (i.e. the constitution of information as value through differential speeds).

Nonetheless, delay also impinges in a more complicated and unexpected way in playing Avara, and these complications are promising. A much younger friend of mine is always urging me to play computer games. Agreeing to try this game with him, something struck me as he quickly won a succession of games. He was not only anticipating most of my movements, and my gestures, he was also anticipating and manipulating in certain ways the delays introduced by the network we were playing on. For a beginner, these delays are incapacitating, for the scene changes before a gesture makes (or more to the point, misses) its mark on the screen. Because gestures have to be passed over the network, and then reintegrated into the BSP tree along with anything another player has done in the meantime, there is perceptible delay between hand and eye. Gestures and images shift in and out of synchronization as real-time processing occurs over the networks. The beginner cannot adequately anticipate the delay between the time of touch and the time of the image in order to coordinate the two. Worse, when opponents come into close quarters, the rate of events (firing and movement) burdens real-time processing even more, so that the screen images slow down just when they should be most responsive.

The technical name for this phenomena, which is well known in real-time applications, is *latency tolerance*. It refers to the degree of lag between an event and the completion of its computational processing; between, for instance, a gesture and what happens on the screen. The remarkable thing about latency is that it can be tolerated. Players habituate themselves to the delay in the circuit between hand and eye and eventually, within certain limits, do not even regard it as an obstacle. Embodied anticipation can 'overcome' the delay, or render it latent, so that delays in the flux of images are not even obvious to the player. Over time, and through repetition, an exterior delay is gradually remapped or integrated within an altered rhythm of movements, so that gesture runs in advance of the technologized image it should be merely responding to. Analogous to the indeterminacies left open by BSP tree algorithms with respect to 'in front of' and 'behind', habituation to delay entails a re-sorting of relations between 'before' and 'after'.

Latency tolerance also has another facet. Playing over the networks from Australia with overseas players, the delays can become so obvious that some players find them intolerable. (US players will usually, for instance, ask the Australians to leave the game because they're slowing everyone else down.) However, it is not only Australians who slow things down. Delay unpredictably arises from the gestures of others who are not even players, from responses that have not been and could not have been anticipated. No matter how tolerant of delay a gesture can become (i.e. no matter how much delay it renders calculable through habit), it cannot fully take into account the changes in synchronization due to the touch of others, especially that of non-players on the image. Their gestures cannot be segregated from mine, and their responses impalpably alter the rhythms of my response. No doubt their presence can be minimized (as, for instance, when Australians are asked to leave the game) but, in this context, other responses always mark the transit of mine; the propagation of others' gestures transforms mine. In that respect, the 'whatever' body's incorporation of delay can never be as complete or seamless as the ideal of real-time simultaneity promises.

TOUCH AND TECHNICAL IMPERFECTION

This re-sorting of temporal relations, without its double-faceted loss of simultaneity, touches on something more constitutive than a state of technical imperfection. What cannot be anticipated as such in any technologization, what cannot be locked into the projective geometry, is the timing of touch. Certainly, every touch can be precisely 'time-tagged' by the computer, but the system that remains open cannot directly take into account the delay time of its own distribution, or of the way in which we corporeally and collectively habituate ourselves to the delays involved in the system. The wavering inconstant anticipation is not susceptible to measurement, since it cannot be known in advance what depth of anticipation has been incorporated into a gesture. Correlatively, we cannot be fully conscious of or in control of the delay that haunts all our gestures, since those gestures are themselves complicated forms of anticipation and response. The system of marks which synthesizes contiguity between bodies and machines is ineradicably open to delay and the effects of anticipations of delay.

Delay permits information to accumulate economic value. Massive technical and economic enterprises seek to render this delay calculable, by investing in information and communication infrastructures which increase the capacity of the networks to accommodate movements of information. Real time, as the temporal horizon of movements of information, seems to entail a loss of temporalization. In an important sense, the constitutive value of delay as non-presence, as trace, for time and other seems to be lost here. In his deconstructive approach to technology, Bernard Stiegler argues that reduction of delay produced in real time specifically challenges the non-coincidence between writing and reading which is constitutive for differing-deferring thought (Stiegler, 1996, 77–8). Because no time is lost, time (as temporality and historicity) is lost. If informatics seeks to render all delays calculable, it can only do so by investing or manipulating the intervals, the sites of differentiation, or constitutive incompleteness within sociotechnical collectives. Real time attempts to collapse the intervals between event and its reception, so that the event is structured by its processing. Those structures utilize a technically managed circulation of information to introduce speed differentials which can then be rapidly and briefly capitalized. Under the 'race condition' production of exchange value, destabilizing events are constantly captured and consumed as spectacle.

It remains to be thought, however, what manner of singularity can inhabit this staging of coincidences between events and structures. If there is play (in the sense that Agamben describes as the manipulation of the human temporality materialized in objects) in computer games, then it is play that somehow must diverge from the economic transformation of events into structures. Presuming that computer games figure in that spectacle as exemplary experiences of real-time capture of delay, I have been asking what scope there is for play to manipulate structures that are already set up to absorb a high rate of contingency. The question is thus, how can there be play when the structures involved are already explicitly organized as buffers for indeterminacy?

Occasionally, computer games manipulate the materialized temporality of information as economic value. It would be possible to interpret the structures of play within a more detailed set of technical, economic and social problems concerning movement, delay, thresholds, fragmentation and exclusion. Here, the focus has been on something more elementary, yet elusive. The discussion has sought to identify certain kinds of indeterminacy in real-time animated games, both in the virtual structuring of images as coded spaces open to gesture (exemplified in the BSP tree) and in the incalculability of delays stemming from the anticipatory element of any gesture. Understanding this materialization as the 'whatever' body entails the step of apprehending the linkages between bodies and images in the game as the incipient ethos of an informatic whatever.

These are perhaps slender supports on which to rest a response to the injunction presented by the notion of the 'whatever' body, an injunction to think how to belong to impropriety, or how to singularly inhabit indifference. 'Whatever is the thing *with all its properties*, none of which, however, constitutes difference. In-difference with respect to properties is what individuates and disseminates singularities', writes Agamben (1993b, 19). How would such a notion of in-difference, in which proper/singular and universal/common change roles and interpenetrate, bring us into contact with an ethos or taking-place associated with information? The interplay between contingency-absorbing structures and the unpredictability of delays is neither completely proper to computer games (or by extension, information more generally) nor simply improper to them. Rather, from time to time play occurs, not under the mastery of any subject, and the line between the incorporation of delay into the informational process and an unanticipated delay wavers. These unpredictable occasions are neither simply technical nor properly non-technical. Time is lost, and the passage of the 'whatever' body between common and proper becomes visible.

NOTES

1. During 1998, according to newspaper reports, one-quarter of the overall revenue received by Sony Corporation came from sales of its PlayStation computer games console. (*Sydney Morning Herald*, no. 84, 20 March 1999, p. 10.)
2. For a history of the personal computer (PC) see Campbell-Kelly and Aspray, 1996.
3. In general terms, moving further into the detail of Agamben's acccount, play renders tangible the differential margin between the 'once' and the 'no longer'. These two terms, the 'once' and the 'no longer' can be taken to refer respectively to the synchronic and diachronic dimensions of social formations. 'At once' refers to the synchronic nature of structures, the 'once upon a time' of myths, present all at once; 'no longer' refers to the diachronic pole when every present moment falls away into the past without being retained or stored. Play, as the margin between them, can be understood in a mechanical sense too, as the amplitude of the movement between them.
4. Again, more recent and complicated games than Avara seek to display every gesture that has an effect on the game scene. Increasingly, every object in the scene bears the marks of previous contacts. All objects in the scene are susceptible to damage, not just the bodies of other players or the enemy-others.
5. For the first academic publication describing the BSP tree, see Fuchs *et al.*, 1980. The patent for the technique is US patent 5274718: Image representation using tree-like structures and can be viewed at www.delphion.com/details?pn=US05740280_.
6. This is Aristotle's definition of *physis* (cf. Aristotle, 1996, 33).

CHAPTER 6

Life, collectives and the pre-vital technicity of biotechnology

Gene mapping is a particular kind of spatialization of the body. . . . How does it get mistaken for a non-tropic thing-in-itself?

Haraway, 1997

Life as information implies complicity with realtime.

Stiegler, 1996

We know that technical mediations can saturate life. Chapter 2 showed that something as ostensibly simple as a stone axe presupposes deep and intimate transformations in a living body. A tool implies a corporealization for someone or something. Recently 'life', from the supermarket tomato, through genetically altered rabbit viruses, to cloned sheep and attempts at human somatic gene therapy, has undergone a fairly literal technological rendering. Life, as we are told constantly, is now being explicitly designed or engineered. No longer individual human bodies, but life as a diverse and intricately overlapping milieu has become an open and disperse engineering site, the object of mapping programs, financial speculation, voracious property claims, and massive state and corporate funding. Numerous warnings about new biomedical and biotechnological practices and the effect they could have on our norms (nature, kinship, family and health) are now taking place. Life has become intensely technological, or at least newly susceptible to an association with the potent but almost empty abstraction, 'technology'. How could such a connection unfold between technology and life? In the terms that I have been developing, what kind of transduction or collective individuation is occurring as life becomes explicitly technological? Something should be becoming clearer. As a universal, the notion of 'technology' (and, by association, of 'biotechnology') has very poor traction in the domain of collective life. It appears more or less as an empty signifier, capable of sliding across different signifieds such as 'progress', 'exploitation', 'freedom' or 'control.' The general problem threaded through much of this book has been how to negotiate a path between an overly general notion of technology and the localized under-represented milieus of technical practices. The localized practices I have focused on have been largely informatic. Why? First, it is hard not to be affected by the hype that has surrounded these technologies for the past few decades. The hype should not be dismissed without consideration. These devices are fascinating because they have allowed the webbing together of practices. Thus, even if certain isolated devices have been fetishized (desktop computers, for instance), it is strongly arguable that informatic-material practices are largely responsible for the existence of large-scale communication *ensembles* such as the Internet.

Earlier discussions of timing regimes and network devices showed that technologies are not unified, discrete, meaningful *or* meaningless in themselves. Although they often figured as discrete devices, they are deeply entwined with collective practices. Similarly, even if biotechnology is currently a fetishized domain, conceptualizing the existence of a technical ensemble entwined with collective life remains a problem. That problem of the ensemble, of how to think about something that is neither a structure nor an individual substance has motivated my recourse to the notion of transduction, with its focus on individuation as the articulation of diverse realities. Talk of technicity, and in this chapter, of the technicity of biotechnology, stems from the need to highlight a formative and radically contingent *collective* entwining between living bodies and information, rather than an interfacing of individual bodies with machines.

Foreshadowing the discussion in this chapter, the immediate problem is how to formulate what is collectively at stake in biotechnology. We do not yet know what life becoming technical means for collective life, but it is clear that we need ways of talking about it as an event.

Part of the problem is the uncomfortable term, 'biotechnology'. It does not sit easily among all the other artefacts and practices we usually understand by technology. In this respect, biotechnology is not just one example among many others. Nor, in terms of the linked series of 'examples' I have discussed in previous chapters, is it an isolated example. It brings the problem of the living body (posed in the context of Judith Butler's theory of corporeal materialization) back to the fore. At the same time, it represents a significant, historically analysable encounter between living bodies and the material practices of information in which we are all implicated, knowingly or not. Even if it might not be possible to say what biotechnology is (and this will be the first problem addressed in this chapter), we can see how problems in defining it flow from something worth thinking about, something which I will conceptualize in terms of the notion of the 'pre-vitality' of collective life. The aim of this chapter – to analyse why biotechnology is difficult to think about – sounds *prima facie* negative. It is not, insofar as an incapacity to fully represent something means that something can happen.

TRANSDUCTION AND PRE-VITALITY

The term 'collective life' has often surfaced in this discussion. It has operated in the context of an explicit distinction between living and non-living entities. From the standpoint of transduction, the basic distinction between the living and non-living can be framed both topologically and temporally. Topologically, the transductive individuation of a non-living thing takes place at a surface or boundary. The surface of a crystal grows in its solution, but the interior layers of the crystal, which have already structured themselves, remain stable and are not altered by further growth. Transduction occurs between the potentials of the solution and inter-atomic forces in the crystal. By contrast, in living processes, both the interior and the exterior of the entity grow (through regeneration), and the entity itself, by virtue of its membership of a group of some kind (colony, community or species), can participate in reproduction. The very existence of interior changes is significant. Life prolongs and complicates its individuation through sexual reproduction, heredity, mobility and communication.

It constantly resolves problems for itself through perception, movement, nutrition, excretion, communication, even by dying. In doing
so, the line between inside and outside divides and shifts constantly.
The non-living has no true interior, because it does not maintain a
disparity or desynchronization within itself. The living, because it
effectively has both an interior milieu and an exterior milieu, dephases
itself. This can be understood as a topological complication. Gilbert
Simondon claims that in the living, 'interiority and exteriority are
everywhere' and they are in contact with each other (Simondon, 1995,
159). Considered from the standpoint of this temporal complication of
topology, life or a living body does not map on to any single surface,
and transductive processes are not located solely on a single surface.
Manifold and differentiated exterior milieus – perceptual, alimentary,
semiotic, energetic, symbiotic – continuously fold into an interior,
which folds itself outwards through reproduction, and growth. Life is
transductive because it is not individual.

In terms of this abstract contrast, technologies occupy a curious
position. Although machines, devices, tools and infrastructures are
identifiably non-living, they belong to living collectives. They lie within
the topology of collective life, yet themselves lack interiority. This
borderline position is well expressed by the meaning of transduction
in contemporary molecular biology: 'In addition to their own DNA,
phages can acquire cell DNA from the bacteria they infect and transfer
that newly acquired DNA into other bacteria in the course of
subsequent infection. This phenomenon is called *transduction*' (Berg and
Singer, 1992, 84).

A phage – a virus that infects bacteria – serves as a carrier or vector
of genes from one cell to another. When a bacteria cell is infected by
the quasi-vital phage, fragments of the bacteria's DNA are incorporated
into the phage DNA. When the altered phage infects other bacterial
cells, the fragments replace corresponding DNA segments in the
recipient cells. The transfer of genetic material is called *transduction*,
and it provides a general model for biotechnological manipulations of
heredity.

Transduction in this sense was also one of the earliest breakthroughs in molecular biology. Understanding and manipulating bacterial transduction was pivotal in the development of recombinant DNA

technologies (Judson, 1992, 55). The phage itself, as a kind of virus, is not strictly alive. We could say that it is *pre-vital*. It wavers along a line between the living and the non-living.

While I am not arguing that technology acts as a virus that infects collectives, I do think that this technical meaning of transduction touches on something important about biotechnology. We cannot ignore the fact that biotechnology gains traction in living processes only by unravelling them in milieus where not everything is alive, and where processes of transfer and mutation proceed on carefully inscribed and circumscribed surfaces (microarrays, gels, blots, chromatograms, etc.) rather than in convoluted, hidden, diverse pathways. Biotechnology actually heightens the experience of complex inseparability between the living and the non-living because of the topological transformations it introduces. Broadly speaking, it brings 'life', whose meaning was already the object of collective historical contestation, into explicit contact with 'technology'. Michel Foucault's notion of 'biopolitics' can be seen as furnishing the background for this point (Foucault, 1978; cf. Agamben, 1998). Summarizing Foucault, Paul Rabinow (1992, 236) argues:

> Historically, practices and discourses of biopower have clustered around two distinct poles: the 'anatomopolitics of the human body', the anchor point and target of disciplinary technologies on the one hand, and a regulatory pole centered on population, with a panoply of strategies concentrating on knowledge, control and welfare, on the other.

Today, that clustering is being redistributed into a distributed ensemble of living and non-living actors. Sequencing robots, databases, immortal cell-lines, stem cells, hybridomas, hybridized mice and radioisotopes are just some of the diverse technical elements of the ensemble. Most accounts of subjectivity, corporeality, history, society, culture and technology assume that something living animates technology. Biotechnology complicates that assumption. It involves a kind of design, and a kind of engineering, but a designing that intimately associates living and non-living elements.

This significant redistribution and gathering of diverse actors in ensembles whose outlines are not yet clear has consequences for the representability of biotechnology. In stock-market speculation on bio-tech companies, in frequently published newspaper, television and journal announcements of the next breakthrough in disease diagnosis and treatment, in genetic modification of crops and farm animals, or in pharmacogenomic product launches, there is something more than unbounded hubris concerning life and its profitable technological manipulation. Like the affect surrounding computers, it flows from a significant reorganization of practices. Conversely, when ecological arguments against the deployment of genetically modified organisms insist on the interconnectedness of life and, in particular, on the unpredictable interactions that occur between different species or communities of living things, they point to something other than a competing scientific discourse (ecology) on life and its complexity. Both the ecological responses to biotechnology and the biotechnophilia of agribusiness and drug companies cannot say why the living and non-living are entwined to such a degree that something called 'biotechnology' could become an object of contention. In other words, without an account of the ways in which the technical and the living are entwined, the duress which biotechnology imposes on living bodies can only be mis-recognized as 'progress' or 'exploita-tion'. The widely divergent responses to biotechnology might be understood as flowing in part from the indeterminate status of this complication of the living and the non-living, which cannot yet be thought, represented or communicated as a coherent or complete phenomenon.

In this chapter, those difficulties in representation will be analysed in terms of three problems. The first is the troubled connection between biotechnology and the notion of life as information. Second, the emergence of a new sub-discipline called 'bioinformatics' concerned with the ordering of massive amounts of biological information will be discussed. Finally, the problem of life as a 'history of error' will be examined.

BIOINFORMATICS: THE COLLAPSE OF MATERIALITY AND METAPHOR

Informatic life and its errors

The first problem has been recognized and thoroughly developed in histories of modern biology: even before the double helix structure of DNA was made visible, life had started to become informatic. Georges Canguilhem writes:

> Insofar as the fundamental concepts of the biochemistry of amino acids and macromolecules are concepts borrowed from information theory, such as code or message; and insofar as the structures of the matter of life are linear structures, the negative of order is inversion, the negative of sequence is confusion, and the substitution of one arrangement for another is error. (1991, 276)

Donna Haraway summarizes: 'much has been written about how the reconstitution of biological explanations and objects of knowledge in terms of code, program, and information since the 1950s has fundamentally recast the organism as a historically specific kind of technological system' (1997, 97). Error, as Canguilhem's observation indicates, is inherent to life and heredity, and this is something that the informatic conception of life as the transmission of hereditary messages renders manageable. The mutation of biology as a research practice is well recognized by contemporary biologists too: information technologies such as the GeneBank database 'have revolutionized biology, providing researchers with powerful tools to hunt for new genes, compare the way genes have evolved in many different organisms and figure out the functions of newly discovered genes' (Pennisi, 1999, 447). Almost from the outset, the presence of information in biology – as metaphor, as technical practice – has been at the same time erroneous and powerfully enabling.

When heredity is understood as a transfer of information, growth and reproduction are driven by permutations and combinations. To quote James Watson and Francis Crick from the 1953 announcement of the structure of DNA: 'In a long molecule, many different permutations are possible, and it therefore seems likely that the precise sequence of the bases is the code which carries the genetical information' (Watson and Crick, 1953, 967). It was thought that sequences of nucleic acids carried instructions coded in permutations of bases. The notion of the genome as a program, and cellular life as a computer that executed the program, gained precedence in the late 1950s and early 1960s. Confusions between life and information were present from the outset. As Evelyn Fox Keller writes:

> With Watson and Crick's invocation of 'genetical information' residing in the nucleic acid sequences of DNA, some notion of information (however metaphorical) assumed a centrality to molecular biology that almost rivalled that of the more technical definition of information in cybernetics. (Fox Keller, 1995, 94)

In the context of molecular biology beginning in the 1950s, the notion of genetical information was understood as a program which issued commands to the living cell in the form of proteins. This understanding of genetic information regarded DNA as the controlling script for life, and metaphorically carried over Erwin Schroedinger's notion of the genetic 'codescript' as formulated in his 1943 lectures 'What is Life?' (see Fox Keller, 1995, 3–42). By contrast, the concept of information presented by Shannon's mathematical theory of communication in 1949 defined information differently: information measures the degree of uncertainty contained in a signal. Genetic information, as Fox Keller points out, was thus somewhat at odds with the cybernetic concept of information. In life, coding errors make all the difference between life and death; coding errors can be fatal. Moreover, genetic information acts as both a legislative code and executive power (p. 95). Information in the genetic code was thought to contain all the instructions for the development of a living organism over time, whereas information as understood by communication theory measured how many possible different messages could be transmitted between a given transmitter and receiver.

Confusion between genetic information and cybernetic information persisted despite its manifest problems: For example, 'as early as 1952, geneticists recognized that the technical definition of information simply could not serve for biological information (because it would assign the same amount of information to the DNA of a functioning organism as to a mutant form)' (p. 19). Perhaps more importantly, it allocated no active role to any other part of the cell or organism, no organizational complexity or dynamism beyond that programmed by inherited genetic instructions. Today, the notion of gene as program continues to attract criticism for the same reasons. Molecular biologist Richard Lewontin criticizes the notion of DNA as program. His criticism of the 'vulgar biology' of genetic determinism (prominent in much of the debate around the Human Genome Project) during the early 1990s argued that DNA has no power to reproduce itself, no activity apart from inherited cellular structures: 'we inherit not only genes made of DNA but an intricate structure of cellular machinery made up of proteins' (Lewontin, 1992, 33). In many respects, these criticisms are directed mainly against the popular conviction that inherited genes impose a kind of biological fate over life.

Whatever the limitations of the conventional so-called 'Central Dogma' view of DNA, which said that information flowed irreversibly from genes to proteins, there is no question of contemporary biology, even in its repudiation of the irreversible flow of information moving from the DNA to proteins, dispensing with informatic metaphors. The Central Dogma has been replaced by a more thorough-going informatic view of life. Over the last few decades, the view that DNA is vitally active, and the rest of the organism is passively organized has been replaced by a much more complicated rapprochement between DNA and extra-nuclear processes. (A single gene can be read out in different ways or in portions that are spliced differently. Protein synthesis is also more complicated than simple encoding.) The current position, especially as represented by the resurgent interest in developmental dynamics, cell differentiation and 'networks' of gene regulation imbricates molecular biology more deeply with distributed processes of communication and exchange. Many examples could be cited here. First of all, the above quote from Lewontin shows that cellular structures are to be understood as 'intricate machinery'. Although there is nothing new here, since intricate mechanism has been a guiding model for life since at least the seventeenth century, today the precise character of the mechanism becomes increasingly informatic.

Whereas 40 years ago only the genetic material was informatic, today the whole cell is a distributed computation: '[A]n alternative metaphor of DNA is data to a parallel computing network embedded in the global geometrical and biochemical structure of the cell' (Atlan and Keppel, 1990, 335).

Again, it is not at all difficult to find words to this effect:

> Genomes of mammals contain a large amount of information . . . which the organism uses during its development and continuing existence. The information stored in this base sequence encodes not only the machinery that each cell requires to carry out tasks like generating and using energy, but also the information for constructing this machinery, the codes used in the construction, as well as the information controlling the selective information readout. (Lehrach *et al.*, 1994, 19)

Or, more recently,

> We're entitled to think of the, let's say, 100,000 genes in a cell as some kind of parallel processing chemical computer in which genes are continuously turning one another on and off in some vastly complex network of interaction. (Kauffman, 2000, 50)

Although the figures have changed (there are now said to be only 30,000 genes in the human genome), the informatic metaphor, rather than falling away, has now clearly extended well beyond the genetic material to include the whole cell, if not the whole organism. Just as imagining computation as a program executed by a computer now falls far short of the complex distribution of computation in contemporary collectives, understanding life as an information system entails taking a far more complicated view of information.

Productive confusion of metaphor and materiality

At the level of descriptive metaphors at least, cross-contamination of cybernetic notions of information and the genetic information strongly pervades contemporary biological understandings of life. In these terms, 'life is constituted and connected by recursive, repeating streams of information' (Haraway, 1997, 134). We can now ask what is at stake in the appearance of this confusion of metaphors and reality. Haraway takes an important step in this direction by regarding the confusion as productive rather than merely erroneous:

> Not only does metaphor become a research program, but also, more fundamentally, the organism is for us an information system and an economic system of a particular kind. For us, that is, those interpellated into this materialized story, the biological world *is* an accumulation strategy in the fruitful collapse of metaphor and materiality that animates technoscience. . . . The collapse of metaphor and materiality is a question not of ideology but of modes of practice among humans and nonhumans that configure the world – materially and semiotically – in terms of some objects and boundaries and not others. (1997, 97)

The existence of competing understandings of information (as command versus measure of uncertainty in a sequence of signals) does not simply indicate irrationality or a failure to think clearly about the ways in which gene action differs from the execution of a computer program. The 'collapse' comes about, and the organism *is* an information system, because of 'modes of practice' that link humans and nonhumans through objects and boundaries that 'materially and semiotically' perform as – and through – information systems. In other words, it may be better to say that the 'error' that mistakes the metaphor of information, programs and computation for the materiality of life may enable something like biotechnology to become a viable object of collective investment. The error that collapses metaphor and practice is an enabling error, in the same way that an error in heredity can be an enabling mutation. Rather than regarding information as an erroneous take on life (as Fox Keller tends to), Haraway argues that the very tissue of the biotechnological hybrid is informatic. Information is not just a metaphor that reduces the complexity of life as an object of biological knowledge, it is also a set of technical-economic practices which trace certain paths and not others.

Bioinformatics and databases as obligatory
passage points

For instance, the accumulation of knowledge of living systems has come to rely heavily on computer databases. Genomic and protein databases are not just repositories of biological data from projects such as the Human Genome Project (HGP), they have become part of the infrastructure of biotechnology. Through several decades of sequencing, these databases have become crucial to contemporary and future biotechnologies. In what ways do databases such as GeneBank enable the organism to become informatic? In the databases, Haraway suggests, 'embodied information with a complex time structure is reduced to a linear code in an archive outside time' (1997, 245). The topological transformation certainly allows different reading and writing practices from those carried out in the folds of nuclear DNA, mRNA and protein synthesis *in vivo*, but those reading and writing practices need to be approached carefully.

The emergence of bioinformatics as a science stems from the problems of sorting and comparing billions of sequences of DNA base pairs. Constant rereading and rewriting of archived linear sequences also focuses on mapping the relation between DNA sequence and the topological structure of proteins. The molecular biologist Walter Gilbert, writing in 1991 about the Human Genome Project, corroborates the point from one angle:

> The problem of working out the human genome can be broken up into three phases requiring inputs that differ by orders of magnitude. First the DNA itself – two meters in length – must be broken into ordered smaller fragments, a process called *physical mapping*. The best estimates of how long the mapping process should take are on the order of a hundred person-years. The second phase – actually determining the sequence of all the base pairs of all the chromosomes – will take three thousand to ten thousand person-years. The third phase – understanding all the genes – will be the problem of biology throughout the next century: about a million person-years, a hundred years of research for the world. (Gilbert, 1991, 85)

The 'input' of *two metres* of DNA presupposes an unravelling of the folded complexity of the cell nucleas into a linear sequence of base pairs. However, can we regard this unravelling as the reduction of information with a complex time structure to 'a linear code in an archive outside time'? Unfortunately, it may pre-empt effective understanding of this transformation to simply say that 'embodied information' (the 6 billion base pairs of human DNA folded into the nuclear of a cell 0.005 millimetres in diameter) is *reduced* to a linear code. The information in the databases is neither simply linear nor the archive outside time. Clearly, the organization of sequence data, and its reading and writing, is very much the object of intensive technical investigation. It constitutes the principal technical problem of bioinformatics. It would perhaps be no more or no less correct to say that the linear writing of a book reduces complex temporal structures of language to linear marks outside time. The transmissions of marks is inevitably accompanied by historically instituted practices of reading and writing which complicate any apparent linearity. On this point, Haraway's account, which provides a sophisticated reading of the informatic materialization of life in contemporary collectives, engages less productively with the topology and temporality of technical mediation. Rather than being outside time, the archive is very much in time in a number of ways. The analysis of genomes has a complex time structure, as indicated by the language of industrial time management – from thousands to a million 'person-years' – scattered through Gilbert's summary. If genomics was 'outside time', then it would not be necessary to relate the mapping to projected completion times and research management plans. Certainly, the recent announcements of the release of the first drafts of a complete map of the human genome have been strongly framed in terms of an historical progression of scientific breakthroughs (Copernicus, Galileo, Newton, Darwin, Einstein). However, the breakthrough involved in completing the sequencing of the human genome is mainly concerned with the large-scale organization of fragments of sequence data in a single database, such as the US National Institute of Health's GeneBank which in July 2000 held 'the sequence data on more than seven billion units of DNA' (Howard, 2000, 47). In mid-1998, completion of the sequence was projected for the end of 2003 (Collins *et al.*, 1998, 683). It was roughly completed at least three years early. There is a reduction in time, but not outside time.

Furthermore, regarded topologically, the reduction is not linear. While the ultimate goal of the genome project is to determine the sequence of base pairs that make up the human genome, this goal can only be achieved by creating an overlapping series of maps of increasing resolution, as Gilbert's comments indicate. Each successive phase of unravelling produces a different kind of map. Each phase also requires an accelerating analysis of the genomic material since the informatic content of the sequence becomes increasingly complex. The process of physical mapping, which breaks down the genome into overlapping fragments, yields only a relatively coarse indication of where genes are located in relation to each other.

Conversely, even having obtained a roughly complete sequence of nucleic acids for a particular organism, it needs to be read and annotated according to complex protocols which are themselves the object of intensive research in the field of bioinformatics. It becomes useful through being linked to the other large-scale databases containing the details of when and where various genes are active, of protein structure and folding, and protein–protein interactions. A sequence becomes more significant when it can be compared to the sequence data of other organisms, ranging from yeast to mice. This work is what Walter Gilbert calls 'the problem of biology throughout the next century'. A simple linear sequence will not produce information about the topological structure or development dynamics of any organism. Highly intensive searching, comparison and sorting of sequences is required before the databases can yield information needed for the sequence-based biology to 'have unprecedented impact and long-lasting value for basic biology, biomedical research, biotechnology, and health care' (Collins, 1998, 682). The linear code will only make sense if read against the codes of other organisms, and against other kinds of maps which permit analysis of the variations in the sequence. The burgeoning field of bioinformatics focuses directly on this problem: how to read the topological and temporal complexity encoded in a linear sequence. A wide and still undetermined array of factors will affect the development of ways of reading genetic information.

The genome and protein databases, along with the computer programs which allow biological research to mutate into database searching and sequence matching, are at the centre of the biotechnological ensemble. More than mere repositories of biological data, they constitute a site of articulation of the living and the non-living. Biologists now say that 'the databases themselves generate new knowledge. The new knowledge is providing a remarkable picture not only of how living systems evolved, but also how they operate' (Doolittle, 1990, 21–2). In general terms (as we will see in a moment), that technical knowledge of 'how living systems . . . operate' flows from the sorting and comparison of sequence data. The archival aspects of biotechnology are highlighted in the genome mapping projects:

> [T]he global Human Genome Project is a multinational, long-term, computative and cooperative, multibillion-dollar (yen, franc, mark, etc.) effort to represent exhaustively – in genetic, physical, and DNA sequence maps – the totality of information in the species genome. The data are all entered into computerized databases, from which information is made available around the world on terms still very much being worked out. Computerized database design is at the leading edge of genomics research. (Haraway, 1997, 246)

The convergence of informatics and biology meshes the techniques used to organize and retrieve scientific knowledge with the concepts used to understand the organism. The practical consequences of conceptualizing DNA as information have turned out to be closely bound to the problem of how to organize and make technical use of that information.

Gene-mapping, as undertaken in the HGP, produces various kinds of maps (genetic linkage maps, cosmid scale physical maps and, ultimately, a sequence of base acids) largely residing on computer databases. There are at least ten different types of genomic maps that must be organized in relation to each other (Cantor, 1994, 2). The different kinds of maps involved in sequencing the human genome overlay each other, producing multilayered inscriptions of the topology of the organism's nuclear DNA. The troping of sequence data as *maps* plays an important part in seeing molecular biology as racing to claim biopolitical territory (Haraway, 1997, 162–3). The diverse technical mediations involved in its mapping projects need to somehow encounter each other on common ground. Again, making that common ground accessible is a central problem for informatics. Through the development of search and comparison techniques, it offers ways to superimpose different maps on each, and to detect homologies and similarities that can be capitalized upon.

Given the focus of this chapter on the problem of saying what biotechnology means for collective life, focusing on databases seems a litte abstruse. But one justification for it comes from Haraway's notion of *corporealization* providing one account of how diverse technical and social realities intersect. That notion forms part of her more general attempt to render visible the technologies that allow semiotics and matter to reversibly transmogrify into each other: '[i]n art, literature and science, my subject is the technology that turns body into story, and vice versa, producing both what can count as real and the witnesses to that reality' (Haraway, 1997, 197). Although she does not settle on any single artefact, the collection she discusses (Robert Boyle's vacuum pump, computer chips, foetuses, laboratory mice, the human brain, seeds, etc.) include genome databases as particularly condensed sites of interconnection of 'knowledge-making practices, industry and commerce, popular culture, social struggles, psychoanalytic formations, bodily histories . . . and more' (1997, 129).

Could it be argued that, together with the protocols for reading and transcribing genetic information between different living and non-living supports, the genomic and protein databases participate in a corporealization? How can something like a genome database be involved in corporealization? Interestingly, Haraway presents corporealization as the mark of the collective in a body:

I am defining corporealization as the interactions of humans and nonhumans in the distributed, heterogeneous work processes of technoscience. . . . The work processes result in material-semiotic bodies – or natural technical objects of knowledge and practice – such as cells, molecules, genes, organisms, viruses, ecosystems, and the like. . . . The bodies are perfectly 'real' and nothing about corporealization is 'merely' fiction. But corporealization is tropic and historically specific at every layer of its tissues. (pp. 141–2)

Corporealization occurs, Haraway argues (invoking Bruno Latour) when an 'obligatory passage point' (p. 164) interacts with and inflects what passes through it. It becomes an interface which diffracts and scatters. It shifts the boundaries, the intelligibility, legibility, visibility, activity and palpability of a body understood as 'simultaneously a historical, natural, technical, discursive, and material entity' (p. 209). Genomic databases are becoming obligatory passage points for a whole range of biomedical and biopolitical projects. They are 'material technologies through which many must pass and in which many visible and invisible actors and agencies cohere' (p. 270), both human and non-human. Haraway writes, 'something peculiar happened to the stable, family-living, Mendelian gene when it passed into a database, where it has more in common with LANDSAT photographs' (pp. 244–5). This is a striking point whose implications are worth developing. The molecular biologist Walter Gilbert describes the 'family-living gene' when he writes: 'The information carried on the DNA, that genetic information passed down from our parents, is the most fundamental property of the body' (1992, 83). His comment marks the genome as the 'substance' of a body. Haraway's point, by contrast, highlights the fact that 'DNA as information' can only become 'the most fundamental property of the body' through mediations that permit DNA to be configured and communicated as information. DNA's status as *information*, and its status as a 'fundamental property of the body', comes to light only through the reading, copying, comparing and sorting of genetic sequences carried out by a host of technical mediations currently circulating through computer databases. The information transmitted by genes only becomes 'fundamental' in the context of genomic corporealization which renders it visible.

A specific and well-known corporealization passing through genomic databases occurred when a patent was granted to the US National Institute of Health over a cell line derived from a Papua New Guinean man (Beardsley, 1996). While considerably more could be said about this case, the particular cell line became commodifiable because its traits were seen as defining different norms of health (in this case, in relation to heart disease). The isolation and appropriation of this singularity is typical. On a global scale, genomic corporealization means managing roughly three million differences per individual human. Charles Cantor, chief scientist of the US Department of Energy's genome project, writes:

> I would like to imagine that our technology will become good enough to enable us to sequence everybody. When you are born you are sequenced, and if we are clever and compress the data down, so that we have a database of differences between people and throw away all the stuff that is the same, and take the extremely conservative estimate that we differ from each other at the level of 0.1%, that means 3 million differences per individual. If we assume an (underestimated) earth population of 4 billion individuals, the resulting database will be required to hold (just handling the differences between humans) information about 10^{16} base pairs. (Cantor, 1994, 17)

By concentrating on genetic information contained in nuclear DNA, the HGP was able, sooner than originally projected, to map and sequence one idealized human genome. Already, SNP (Single Nucleotide Polymorphism) databases are supplementing this ideal genome by mapping some of the tiny genetic variations between individuals. Bioinformatic corporealization focuses on these differences, on what Cantor predicts will be implemented as 'a database of differences'.

Corporealization as a distributed process

Every technical mediation entails some kind of corporealization. The problem is how to delineate that corporealization in the context of technological ensembles. Their technicity, and the way they are interlaced with collectives, is not easy to grasp. The technicity of an ensemble refers to the transductive process of individuation that emerges from interaction between its technical elements. Chapters 2, 4 and 5 have made that point mainly in the context of information networks. Biotechnology targets bodies in the biopolitical domains of health and nutrition. In general terms, what Haraway understands as 'the distributed, heterogeneous work processes' supporting biopolitical corporealization can also be explained in terms of the practices involved in contemporary biotechnology. Databases are one place where a specific technicity which links together different informatic and living processes becomes more legible.

For instance, in general terms, the practices associated with the sequences stored in the databases can be grouped under:

(a) *Hybridization*: heredity information can be induced to move between different organisms and, today, chemical substrates. (This is the way that molecular biology uses the term 'transduction'.) This process of *hybridization* is fundamental to biotechnology, contemporary and past. It has a both more general and specific meaning. In general terms, hybridization means the process of producing new organisms from the combination of existing organisms. For instance, the domestication of grain crops in paleolithic agriculture is thought to have occurred via the accidental transfer of genes between different species of grass. Human culture has long relied on the results of accidental transgenic movements or hybridization in food plants ('Chromosomes show that modern wheat began when two species of grass hybridized' (Jones, 1994, 269)), or between animals (donkey + horse = mule).

Hybridization is also a basic technique used to analyse genetic material. A number of kinds of hybridization are vital to the processes of mapping genes and manipulating their expression. As we saw earlier, a specific kind of hybridization called 'transduction' carried out by viruses between bacteria allowed geneticists in the 1950s to move small selected portions of genetic material into bacteria (Judson, 1992, 55). The laboratory capture of transduction as a technical process enabled genetic modification of other organisms. In its general sense, hybridization permits the production of somatic cell hybrids (e.g. mouse–human cell hybrids) which have been very important in the industrial-scientific mapping of human genes. In its specific sense used by molecular biologists, 'DNA hybridization means the process of "annealing" [a term from metallurgical craft] strands of DNA or RNA with each other. The technologies of "hybridization micro-arrays" or "gene chips" for instance are currently inducing a quantum leap in the speed of analysis of genetic information since they open the interaction of gene complexes within a living cell to analysis' (Marshall, 1999, 444).

(b) *Selective replication of genetic material*: the primary hereditary substance or genome can be manipulated as a linear sequence of coded information only if it can be isolated in sufficient quantities. Genomics relies on techniques for extracting, copying and sorting genes as sequences of nucleic acids, and for reading and writing proteins as sequences of amino acids. Beginning in the early 1970s, a stream of technical innovations based on the specific catalytic activities of certain enzymes (polymerases, transcriptases, restriction enzymes) allowed sequences of nucleic acids (DNA and RNA) and amino acids (composing proteins) to be selectively copied and edited. The reading and writing of selected portions of hereditary information have tended to confirm the status of the inheritance as an archival system or individual life as the transmission of information. The most prominent example of such a technique is probably PCR (Polymerase Chain Reaction). In the late 1980s, the now heavily used PCR was developed at Cetus Corporation (Rabinow, 1997). It allows large numbers of copies of any given sequence of DNA to be readily synthesized *in vitro*. Once arbitrary amounts of genetic material can be copied independently of their containment in the nucleus of a living cell, genetic information has been effectively detached from its embodiment in species or individuals.

The combination of the two previous modes of practice – hybridization and selective replication – generates the huge volumes of genetic and protein sequence information currently reproduced, sorted and transmitted in and between many different computer databases. These databases then serve as the substrate of new drug research (Haseltine, 1997). The important point here is not to say exactly what role the databases play in contemporary biotechnological research (although this definitely needs a lot more concrete analysis). Rather, I am suggesting that whatever shape biotechnology is taking stems in part from the way in which information is organized in the databases. Biotechnology as it moves out of laboratories into fields, clinics and factories is linked ensemble-wise to the databases. This linking is complex and highly mediated. As mentioned above, a whole series of maps and translations moves across living and non-living substrates before something as refined as sequence data can be stored in a database. Contemporary biotechnology is absolutely reliant on the storage, processing and retrieval of large quantities of genetic information in computer databases. Standing at the confluence of techniques of hybridization and replication, the computational sorting, comparison and matching sequence data is central if the genetic differences between bodies are to become visible and manipulable within the life of the collective.

The metastability of a non-living–living ensemble

The databases, along with the other 'obligatory passage-points' Haraway refers to, such as foetus, chip, brain and virus, can be understood as markers of disparity. They partially articulate or join diverse realities. They constitute elements of a technological ensemble at which the potentials for divergence and further restructuring are supersaturated and metastable. Topologically, they stand at points where an individuation of the collective is most likely to be triggered. They attest to a collective individuation. Gilles Deleuze, echoing Simondon, writes, 'individuation presupposes a prior metastable state – in other words, the existence of a "disparateness" such as at least two orders of magnitude or two scales of heterogeneous reality between which potentials are distributed' (Deleuze, 1994, 246). While the notion of corporealization points to the ensemble of connections between living and non-living entities which envelops biotechnology, the notion of transductive individuation shifts the emphasis on to the metastability from which corporealization stems. What kind of singularities exist in this metastable state, and how are they brought into communication through the articulating processes of transduction?

To understand what kind of individuation and what kind of metastability a contemporary collective might be articulating through biotechnology, the *technicity* of this ensemble needs to be considered. Again, as discussed in earlier chapters, the technicity of an ensemble is not an isolated property, or a hypostatized reality. On the contrary, it is that aspect of a technical mediation which intimately links different milieus together. It structures and sets in play a constellation of forces.

Whatever technological objects compose the biotechnological ensemble – databases, gene banks, clone libraries, microarrays, stem cells, transgenic organisms, etc. – the technicity of the ensemble is not contained in any particular object. It incorporates divergent realities into a structure which exists in a provisional equilibrium. 'Technicity,' Simondon writes, 'is not an hierarchical reality; it exists wholly in the elements, and propagates itself transductively in the technological individual and the ensembles' (1989a, 81). The technicity of the ensemble stems from its elements, but an element of the ensemble, such as a genomic database, does not contain the technicity of the ensemble to which it belongs.

An obligatory passage-point: the technicity of the sequence databases

Specific living and non-living 'material-semiotic' bodies emerge from genomic biotechnology: transgenic organisms (plants, bacteria, animals), bio-drugs, and diagnostic or screening tests which can be highly significant in individual lives (for instance, the technique of pre-genetic diagnosis coupled with IVF (in-vitro fertilization)). Biotechnological objects in their current forms, whether 'rationally designed drugs' (Haseltine, 1997, 82), genetically modified bacteria, goats or cows used for 'pharming' custom-designed drugs, genetically modified crops and food products, prospective somatic or germ-line therapies for disease, or genetically modified bacteria for breaking down crude oil, draw on an ensemble of technical elements, both living and non-living. These include electrophoresis gels, immortal cell-lines, stem cells, sequencing robots, PCR assays, DNA microarrays, etc. Biotechnology as an ensemble consists of living and non-living elements. The ensemble transduces, replicates and sorts certain elements of selected living things via practices which themselves are not fully alive: assays, reactions and, the example I will focus on, algorithms. If we want to understand something of what is at stake in biotechnology for collective life, we need to account first of all for the ways in which biotechnology renders at least these divergent realities compatible. Why is something like a database, or a sorting algorithm, an obligatory passage-point for biotechnology?

The bioinformatic ensemble arises in the tensions and complications between living and non-living entities. While living entities define the general horizon of biomedical and biotechnological knowledges and experimental practices, the general tendency in recent biotechnology, as Paul Rabinow remarks in his account of the technique of PCR, has been to reduce the dependency of experimental work on living systems (Rabinow, 1997, 1). PCR, for instance, makes selected DNA sequences abundant without relying on living systems to reproduce them. So, too, the hybridization, reading and sorting of genetic material in biotechnological practice relies more and more on non-living substrates and techniques such as radioactive labelling, gel electrophoresis, radiographs, DNA array chips, genomic and protein sequence databases, sequence analysis, comparison algorithms and so on. No doubt certain processes, such as cell hybridization, target living cells (bacteria, yeasts, mouse cells) or in the case of 'pharming', complex multicellular organisms such as sheep, pigs or goats, but in general the reduced dependency of biotechnology on living systems is significant. Biotechnology is viable only *in* or *as* this complicated tension between living and non-living processes. It should not be seen as a simple reduction of the living to the non-living, but as a specific co-implication of the living and the non-living. Every techno-logical ensemble brings together living and non-living bodies. There is no purely non-living technological system. However, the bio-technology is distinctive in the way that it configures living bodies as reservoirs of technical elements. The technicity of biotechnology resides within the linkages established between the non-living and the living.

Protein and gene sequence databases occupy a domain that seems remote from the ostensible target of biotechnology, life itself in all its various ecosystems. The database infrastructure is an attempt to cope with the sheer magnitude and variety of interactions occurring within living processes. Because the database contains nothing living, and exists as a set of computer programs and stored data, accessed principally through computer networks, it might seem a strange place to look for the technicity of biotechnology as an ensemble. Databases do not apparently impinge directly on living bodies, in the way that, for instance, a synthetic growth hormone does. The databases (and there are many of them, mostly publically accessible online: GeneBank, the European Molecular Biology Laboratory; DNA Database of Japan, and more recently, the PubMed Project of the US National Institute of Health) are not technical elements in the ensemble; they are really sub-ensembles in their own right. Nevertheless, their remoteness from living bodies belies their importance. The organization of genetic information in the databases implies a specific topological and temporal structuring of the living–non-living ensemble of biotechnology. In other words, the sub-ensemble can illustrate the general point about the transductive articulation of the living and the non-living in bio-technology.

The genomic and protein databases, rapidly growing and incomplete as they are, contain a transcription of the hereditary information transmitted and expressed by life in various forms (human, mouse, rat, dog, zebrafish, wheat, rice, etc.). Genomic databases are con-cerned mainly with the information transmitted when organisms reproduce, and protein databases are concerned with the way in which proteins are structured and function in organisms. Within the organ-ism, genetic information is characterized by repeated, overlapping operations on linear sequences, whereas protein structure, although it is composed of long sequences of amino acids, is characterized topologically, in terms of folds. The topology of a protein determines its specific structural or enzymatic (i.e. catalytic) function in the organism. In short, the specificity of genes is their sequence, whereas the specificity of proteins is their topological folding. Although there is a correlation between the linear sequence of genetic information and the topological structure of proteins (since genes 'code' proteins), defining this correlation is part of 'the work of the next [this] century' (Gilbert, 1991). Both kinds of sequence information (genomic or protein) display hereditary variation. Isolating, sorting and classifying these variations is of crucial biotechnological importance.

In a deep sense, both major kinds of database are in effect concerned with life's history, with the way in which it involves both conservation and innovation. The databases respond to a basic problem whose dimensions are only becoming more apparent now that complete genome maps are becoming available.

The problem is that life, when it 'transmits messages' as sequences which are then expressed in topologically and temporally complicated living systems, observes no general rules. Heredity refers to both the transmission of characters and the expression of characters in an organism (Judson, 1992, 38). As Canguilhem writes, 'heredity is the modern name of substance' (Canguilhem, 1991, 280). When they are materially configured as informatic processes, these two aspects of heredity – transmission and expression – suggest that life can be handled as a set of archiving and reading practices. Quoting Canguilhem again, 'life has always done – without writing, long before writing even existed – what humans have sought to do with engraving, writing and printing, namely, to transmit messages' (Canguilhem, 1994, 317). The first protein sequence elicited (insulin in the late 1940s) showed that 'the sequence of amino acids is entirely and uniquely specified. No general law, no physical or chemical sequence, governs their assembly' (Judson, 1992, 53). At the same time, heredity also demonstrates strongly conservative or cumulative tendencies. If it is possible, for instance, to use rats and mice to model the metabolic processes associated with obesity in humans (O'Brien and Menotti-Raymond, 1999, 459), it is because many gene sequences between mammalian species are homologous or conserved across species. Similarly, there are 'unexpected levels of conservation of gene content and gene orders over millions of years of evolution within grasses, crucifers, legumes, some trees' (Gale and Devos, 1998, 656). The implications of this combination of a lack of generality, regularity or lawfulness in protein and DNA sequences with evolutionary or heredity conservation of sequences are manifold and central to the very possibility of biotechnology. Heredity is an accumulation of accidents, errors and mutations generated in the context of continual rereading. Accident, divergence and error constitute the history of life because life continues to individuate. The variety of life from the standpoint of comparative genetics stems from the spectrum of those accumulated, reproduced errors. In turn, the databases exist as a way of coping with the dimensions of those errors.

INSERTION POINT BY POINT

The impetus to 'map' the entire human genome flows from this lack of regularity. Indeed, genomic maps depend, as well as capitalize, on irregularity. As Judson writes, 'from the start the genetic map of any species-molds, flies, maize or humans-has primarily been the map of defects' (Judson, 1992, 47). If genes, whatever their function (structural or regulatory) operated according to consistent, predictable mechanisms, and were always conserved faithfully, then it would not be possible to analyse their function. Only variations or disruptions in gene action allow biologists to identify them as units of heredity. In any case, if genomic sequences followed general laws, the mapping effort would be obviated. It would be possible to work with general formulae and models to design interventions into organisms.

There are a number of different ways to illustrate the contingencies of heredity, and how it promotes the treatment of life as an information system. First, the concept of the gene as a constant unit of heredity is under serious threat. William Gelbart writes that 'the realities of genome organization are much more complex than can be accommodated in the classical gene concept. Genes reside within one another, share some of their DNA sequences, are transcribed and spliced in complex patterns, and can overlap in function with other genes of the same sequence families' (Gelbart, 1998, 660). This means that even the notion of genes as the components of heredity, as coding for inherited traits, must now be replaced by specific spatial and temporal patterns of interaction between RNA and protein. Philip Kitcher concludes that 'it is hard to see what would be lost by dropping talk of genes from molecular biology and simply discussing the properties of various interesting regions of nucleic acid' (Kitcher, 1992, 130). In effect, this happens in bioinformatics, where sequences are treated without reference to units of heredity, and solely in comparison with other sequences.

Second, specific variations in sequences of nucleic acids in the DNA of individuals need to be taken into account if disease is to be understood and treated genomically. Biologist Eric Lander writes that 'the human genome will need to be sequenced only once, but it will be re-sequenced thousands of times in order, for example, to unravel

the polygenic factors underlying human susceptibilities and predisposi-
tions' (Lander, 1996, 537). Insertions, deletions and 'point mutations'
or substitutions of nucleic acids commonly occur, and continue to
occur throughout the life of an organism, even simply through
exposure to cosmic radiation. The ultimate promise of complete
sequencing of every individual is in fact justified by the individually
varying sequences of proteins and genes in living organisms. The very
idea that it would be desirable to sequence an individual and establish
a database of differences from a normal genome shows that life is being
targeted in terms of individual specificity. As mentioned already,
resequencing of selected portions of the human genome to deal with
individual variations is already under way: 'sequencing selected regions
of the genome in a large number of individuals to identify polymorphisms, which should enable identification of genetic traits associated
with human disease, is seen as an important derivative of the human
genome sequencing project' (Rogers, 1999, 429). The goal of the SNP
(Single Nucleotide Polymorphisms) Consortium is to describe variations of single nucleic acids within selected genes which alter the
activity of the encoded protein. Such variations can affect the way
individuals respond to medications, for instance (see Evans and Relling,
1999).

Third, even if an accurate sequence for a particular gene or protein
is available, no authoritative exegesis of its function can be guaranteed
on the basis of homology with other genes or proteins. The function
crucially depends on the folded topology of the protein coded by the
gene (and even that topology shifts according to context). This
observation is repeatedly made in assessments of the contemporary
state of gene technologies. Automated sequencing has flooded databases
with sequence information, but the work of understanding the function
of the genes described by these sequences has scarcely begun. There
are fundamental problems in linking gene sequence and gene function,
for example 'one of these is the fact that a single protein may have
multiple forms and functions that are context-dependent and that can
never be fully understood by sequence analysis alone' (Boguski, 1999,
454). That is, organisms are 'multiform, intricate and elaborate
physical systems with their operational and regulatory parts assembled
by a series of evolutionary contingencies' (p. 455). When they code

proteins, the combinatorial possibilities of a linear sequence of DNA has to be transposed into the topological complexities of a protein.

Finally, only in a few cases do genes act singly to cause disease. Monogenetic disorders or defects are the exception. The timing of gene action is paramount. Faced with a complex network of interactions between genes, in which triggering, coupling and timing signals are central, there is again no simple way to predict what course of events will unfold within an organism without tracing the complex interactions through every specific step. As Lander suggests, 'the greatest challenge will be to decipher the logical circuitry controlling entire developmental or response pathways' (Lander, 1996, 538). Since the early 1990s, DNA microarray technology has begun to provide a way of mapping these interactions *en masse*. It allows the temporal patterns of gene expression to be elicited. For instance, a cluster of genes that may be inactive early in the development of a strawberry may be relatively active when the fruit is ripe (Marshall, 1999, 445). Drawing on sequence data held in genome databases, an array of synthetically produced DNA sequences ('probes') is laid on a wafer or glass slide. However, even microarray snapshots of clusters of gene activity only sketch an outline of the temporal dynamics of the organism.

In each of these four respects, there can be no abstract representation of the variability of heredity (let alone the differences and changes induced by environmental and developmental conditions) or the functioning of the genome. There can only be point by point mapping of the variations in sequences, the variations in linkages between DNA sequence and protein function, and the developmental dynamics of the organism which appear as topological contact between interiority and exteriority. The possibility of bringing various living processes into the biotechnological ensemble relies on a point by point interlacing of the ensemble with the living systems.

DATABASES AND THE TOPOLOGY OF THE LIVING

The databases owe their existence to a mixture of pattern and irregularity. The point by point insertion of living organisms into a technological ensemble must pass through databases because of this drastically contingent ordering of life as information. As William Gelbart acknowledges, 'were we to be able to read the genomic instruction manual in the same way we can read a book written in a language we understand, we might not need a huge support system of scientific databases' (1998, 659). What prevents us from reading genomes like a book is the fact that life turns out to involve a process of perpetual misreading. Error, misprint, mistranslation, unacknowledged citation, transcription actually constitute the text of life in its reproductive and evolutionary viability.

The databases address this constant misreading by sorting and comparing sequences. In Leroy Hood's words:

> We must be able to extract from a sequence . . . a variety of information, including the boundaries of genes, the presence of regulatory elements, and the presence of sequences that may relate to specialized chromosomal functions such as replication, compaction, and segregation. The key to extracting this information is the ability to compare this sequence against all preexisting sequences to test for similarities. (1992, 147)

The most elementary problem for bioinformatics is how to compare two sequences, and in particular, how to find out whether one sequence is contained in, or could be derived from the other: 'sequence comparison is the most important primitive operation in computational biology, serving as a basis for many other, more complex manipulations' (Setubal and Meidanis, 1997, 47). The so-called 'edit distance' specifies how many insertions, deletions and substitutions of base acids are needed to transform one sequence into another designated sequence. Such comparison has manifold uses, ranging from comparing the genomes of different species in order to construct an evolutionary tree, to deducing the folded topology of proteins, to the forensic identification of DNA. For instance, in order to compare two DNA sequences, GATTACGATTAGC and AATTAC-GATAGC, they can be aligned as follows

GATTACGATTAGC
AATTACGAT-AGC

by inserting a space into the second sequence. The only difference is in the first base (G versus A). The comparison can be carried out on paper by hand and eye while the sequences are short, but it quickly becomes impossible for sequences of biological interest because the number of different possible alignments increases exponentially. This is a case where reading the linear DNA sequences stored in the databases themselves entails temporal complexities which must be addressed by algorithmic treatment. If the databases, or the software which accesses them, treated sequences simply as sequences, then the problem of matching sequences would be intractable because of the exponential growth in the number of different possible alignments. The combinatorial slow-down can only be avoided by using an algorithm which speeds up the comparison. It re-maps the linear sequences on to a different data structure, a two-dimensional array, and breaks down the act of comparing the sequences into comparing first the smallest, and then gradually longer prefixes of the sequences, always using the results of previous comparisons to shorten the work of comparing the longer sequences. To obtain the acceleration in comparison, the order in which the prefixes of the sequences are compared is critical.

Only by such detours and careful ordering of comparison can bioinformatics, and biotechnology more generally, effectively carry out the point by point insertion of living organisms into a technical ensemble. The contingencies and irregularities of heredity mean that life can only become technical via a point by point mapping and comparison of life as a set of transmitted messages. The motivation for the various genome sequencing projects derives from this. Biologists such as Eric Lander and Walter Gelbart would not need to speak of sequencing every person if genetic inheritances were not variable. Genomic databases, and the associated algorithms for sorting, comparing and matching sequences, form only one element or sub-ensemble in the complex, constantly shifting field of contemporary biotechnology. But they are an element concerned with linking various life forms in different configurations by treating them as a set of transmitted messages to be re-read, sorted, cited and combined. They are obligatory passage-points for biotechnology, albeit highly complex and labyrinthine ones, as it reconfigures collective life through recourse to non-human life.

In their current incompleteness, the genomic databases remain somewhat 'abstract'. The technicity of a technical ensemble can be understood in Simondon's terms as the degree of *concretization* attained in the relations between elements of which it is composed: 'technicity is the degree of concretization of the object' (Simondon, 1989a, 72). The process of concretization melds and grafts different elements together. Elements become difficult to disentangle, because they begin to overlap. Boundaries blur between one part of the ensemble and others, and milieus become associated with the ensemble. This process is occurring in the biotechnological ensemble in many different ways. Not only is the work of sequencing ongoing (especially given the goal of sequencing individual variations), but the protocols for reading useful information from the databases are, as some of the earlier comments suggest, still ill-defined. Despite their sophistication and reliance on the latest and most powerful information technologies (e.g. Celera Genomics, the company that sequenced the human genome in competition with the HGP, used supercomputers to order and sort genetic sequences), the *technicity* of these databases is relatively low compared to what it might become. The transductive processes at work in biotechnology have begun to interlace databases with diverse life forms via a panoply of techniques of hybridization and analysis. Point by point, a technological ensemble enmeshes itself with 'target' organisms. While life becomes technological only in a fine-grained assembly of elements, genomic DNA forms a privileged core around which, or through which, the technicity of biotechnology incorporates or corporealizes the living.

Why is it so difficult to collectively represent what is at stake in biotechnology? A short answer would be that whenever the limits of a collective are contested, slippages and failures in signification are triggered. A more involved answer focuses on why it is hard to think through the interleaving of specific technical processes within collective life. Collective life is not fully alive. Canguilhem wrote that 'in order to understand living things one needs a non-metric theory of space, a science of order, a topology; one needs a non-numerical calculus, a combinatorics, a statistical machinery' (1994, 317). In order to manipulate living things on the basis of such an understanding, one needs a machinery of ordering and comparison. The genomic and protein databases represent a non-metric, statistical topology of the living. They must be statistical because there is no general set of laws for life other than those contained in the heredity and environmental processes of recombination and mutation through which biological variations arise. They are topological because they address the problem of how a linear sequence can be mapped on to a complicated temporal and topological organization. The complexity of the databases is, perhaps, a good index of the depth at which life is being reorganized. Again, faced with a lack of universal order, the bioinformatic solution is to use sequence similarities to predict structural or topological similarities in living organisms.

Viewing biotechnology as a symptom of collective individuation responds to that difficulty by showing that the individuation occurs as a temporal and topological folding of many different elements, living and non-living. In relation to biotechnology, the major point that I have been developing is this: to accelerate living processes is also to suspend them at some points and, at those points, to detour into the non-living. Debates over the significance of biotechnology generally overlook the implications of this point, and range between two poles. At one pole, biotechnology is said to change nothing, since humans have always cultivated and selectively bred living organisms. It is often argued that 'for millennia, humans have successfully modified the genetic makeup of organisms through selective breeding' (Berg and Singer, 1992, 221). Countering this argument, biotechnology is said to change everything because now, for the first time, a culture begins to breed and select life forms (human and non-human) without regard for the web of ecological, cultural and economic connections that links living things together. The speed of this selection ignores its com-plications. One side of the debate sees biotechnology as understandable and controllable along the same lines that human societies have followed for millennia. The other side argues that life's interactions or properties cannot be fully understood or controlled in advance, and therefore sees biotechnology as intrinsically risky interference. One side tends to place the technical in control of the living, the other tends to see life as eluding technical control. Ecological arguments against biotechnology deeply oppose the idea (usually promoted by biotechnology corporations or researchers working on patentable research) of limited or precisely controlled interventions in a target organism. For the ecological standpoint (also for dissenting voices within molecular biology itself), unintended interactions between the engineered organisms and the ecosystems they are part of (through growth, reproduction and hybridization) pose a serious risk to living diversity. Ethical, political and moral questions about biotechnology stem partly from ecological understandings of the relational complexity of living organisms. Without wanting to prejudge this debate (which I have risked caricaturing in discussing so briefly), I think both poles share a symmetrical commitment to occluding the role of technical mediations in collective life.

The terms of the debate shift in a transductive perspective. The constituted technical objects of biotechnology – the databases, the genetically modified organisms, rationally designed drugs, statistical screening and forensic test-kits – attest to a different folding of living bodies and collectives. As an ensemble, we have seen that it links the living and the non-living point by point through informatic modes of ordering. Information is not merely a mistaken metaphor. It is an ongoing enabling error. As a consequence, certain rhythms and patterns of reproduction, growth and death of selected life forms are suspended. Clearly the web of connections between and within living things is interrupted and reorganized. The interruptions and suspensions are localized and specific. But so too are the limits, the topology and temporality of our collectives; that is, what or who we mean by 'humans' transmutes. The argument that biotechnology is nothing new ignores the metastability of a collective that lives in part through the non-living. The singularity of biotechnology crystallizes at least in part from the informatic linkages between living and non-living entities. These linkages are historically instituted within and by a collective.

Everything is indeed connected, as the ecological standpoint argues. Yet biotechnology would be a mere abstraction if those connections were not metastable. Arguments about the commodification of seed stocks, and exploitative acquisition of property rights over various elements of living things, as set out in much recent critical work on biotechnology (Shiva, 1995; Fowler, 1995), cannot be ignored. However, what crucially distinguishes biotechnological processes from other living processes is the historically specific suspension of life, growth, reproduction and death within the context of a human–non-human collective. Biotechnology is not a simple interference in living processes. It occurs as a singular conjunction of processes that conjoin the living and the non-living. Attending to how the living and the non-living are grafted on to each other may prove to be a more flexible and adaptive response than simply collapsing the biotechnological on to the domain of the living.

Conclusion

remind why something similar [handwritten margin note]

In the last pages of *French DNA: Trouble in Purgatory*, in a formulation reminiscent of Foucault's concept of 'eventalization' (Foucault, 1991), the anthropologist Paul Rabinow writes: 'from time to time, and always in time, new forms emerge that catalyze previously existing actors, things, temporalities, or spatialities into a new mode of existence, a new assemblage, one that makes things work in a different manner and produces and instantiates new capacities' (Rabinow, 1999, 180). The concept of transduction can be understood as one way of articulating such singular occurrences. In the context of recent technical mediations, it does so by pointing to the productive tension that couples human collectives and non-human forces, currently those of silicon and 'genetic components'. It focuses on the folding of different forces and elements together as collectives individuate. As a way of thinking about an encounter between divergent realities, transduction helps resist the temptation to explain contemporary technology via one of two major rival principles: that of 'technology' as an empty, metaphysical-ideological abstraction disguising social processes, and that of technology as an ahistorical hand colonizing human cultures with its material structures and logic. Technical practices neither form culture nor are they formed by culture. The transductions discussed in this book can be seen in the barest terms as eventful articulations between realities on different temporal and corporeal scales.

This book has tried to inhabit the tension between 'technology' as a signifier within cultural formations, and technologies when they almost unintelligibly interweave signs and things, living and non-living bodies. The small girl in the museum (see the Preface), as she encounters 'boring old space', apprehends something important about technology. It concerns a global-scale ideological production of high technology as a symbol of masculine subjectivity and superpower nation-states. Modern military technology characteristically symbolizes the sovereign power of states. Its missiles, ships and satellites aesthetically frame masculine executive authority. Technologies, especially weapon systems, obviously act as grand symbols of sovereignty (Nancy, 1993, 43). More generally, technologies such as mobile phones, computers and transgenic organisms have meaning within symbolic systems. They participate in the signification of many things including gender, class and ethnicity. However, technologies also articulate non-living and living potentials together, as we have seen a number of times, in diffuse ways that are not always well represented within existing semiotic systems. Flexible and surprisingly contingent collective structures are involved, as we saw, for instance, in the case of Avara or *Ping Body*. Above all, technologies overflow their role as signifiers.

BODIES AND TIME AS COLLECTIVE LIMITS

Why theorize this tension between technology as meaning and technology as event in terms of bodies and time? Previous chapters have answered this question at different levels. There is first of all a general deconstructive motivation which is linked to the notion of radical contingency and the absence of ultimate foundations for thought. Time and body have been important in recent European thought precisely because they contest the prerogative of consciousness, language and culture to give meaning to things. Although they are not usually mentioned in the same breath, corporeality and temporality share something as concepts: they are difficult, perhaps impossible, to think of as such. Both have often been understood in the history of Western philosophy as inimical to thought and meaning. Bodies, in the inconstancy of their perceptions and their ineluctable mortality, contaminate thought with deceptive appearances, error and finitude.

By contrast, a deconstructive response to corporeality says, for instance, that 'if one really thinks of the body as such, there is no possible outline of the body as such' (Spivak and Rooney, 1994, 177). (We saw some of the reasons for this in Chapter 1.) From this position, the body can only be thought of as a contested limit term, and cannot be comprehended directly in itself. The 'extreme ecological position' Spivak adopts on this question highlights the impossibility of topologically mapping or delineating bodily surfaces in themselves. Constant differentiations, exchanges, involutions and projections prevent any such outline from stabilizing except as a particular corporealization or materialization. The concepts of corporeality developed in anti-essentialist feminism (Grosz, Haraway, Butler) have worked against any idea of bodies as natural, ahistorical or even anthropologically constant. The prototypical unthinkability of the body has been crafted into a methodological sieve which allows the historical contingencies of living bodies to be gleaned without immediate reduction to any universal substrate, biological or social. Similarly, time, with its refusal to stabilize as an entity or relation, has long vexed reason with paradoxes of change and non-change. Largely following Heidegger, deconstructive thought has capitalized on just this resistance to meaning to develop a mode of thought that affirms, rather than questions, its own limits. In a deconstructive light, neither time nor bodies can be thought as such. They do not exist as simple entities or substances of which we are conscious or which we directly experience. There is no flux of time of which we could be conscious, only processes of temporalization. There is no body whose outline, substance or form we can be certain of, only processes of corporealization or materialization. *Transductions* borrows from deconstruction (and in particular, from feminist deconstructive theory) an affirmation of the corporeal and temporal finitude of thought.

Aside from this deconstructively motivated attention to radical contingency, there is another direct motivation to develop an account of technology in terms of temporality and corporeality. Our bodies and lives are marked by tensions between different stories, and practices, between different temporalities and topologies. From a transductive standpoint, 'the human' *and* 'the non-human' are the provisional outcomes of a collective individuation in progress. If 'the human form is as unknown to us as the nonhuman,' as Bruno Latour writes (1996, 227), it is because both forms emerge and recede constantly as facets of the same ongoing process of individuation. In Latour's words, 'humans and nonhumans take on form by redistributing the competences and performances of the multitude of actors that they hold on to and that hold on to them' (p. 225). The examples discussed in this book have sought to emphasize the difference between the human and the non-human as an historically contingent and shifting effect of their interlacing. A problematic tension between the human and the non-human, the living and the non-living, flows through these examples. In regard to technology, this tension is heightened because large swathes of philosophical, historical, anthropological, psychological, economic, sociological, governmental, corporate and popular discourse insist that 'the human' is a last instance, something whose stable value and meaning must ground all explanations and judgements concerning technical practices. This means that there is a strong conflict between how 'technology' is figured and how it is embodied in material practices. If we insist, for instance, on viewing technologies as instruments of societies, cultures or civilizations, we lose any possibility of seeing how the capacities and fabric of our collectives are constituted.

THE DOUBLE BIND AGAIN

What is the point of being mindful of that tension or metastability between technology as a grand signifier and a plurality of material-technical practices? The reasons are manifold. As I have said, a growing ensemble of institutions, discourses and knowledges springs from that tension. The fluctuating fortunes of various technologies on the stock markets in recent years, and the unremitting pace of innovation in certain technical domains (such as biotechnology and information technology) foster an intense sense of urgency associated with techno-logical innovations. That sense of urgency pervades stories told about technology. It strongly drives many responses to technological speed. At the same time, new forms of commodification (for example, the proliferation of communication services), diverse fashions and styles (gadget-based lifestyles), music, visual and literary texts, and techno-scientific practices proliferate localized micro-practices indissociably linked with technical ensembles. It is possible, and more relevant than ever, to analyse these astonishingly diverse phenomena from sociolo-gical, historical or anthropological standpoints. They need to be historically and culturally situated. Amidst all this, however, as Donna Haraway (1997, 39) writes, 'the point is to learn to remember that we might have been otherwise and might yet be, as a matter of embodied fact'. 'Technology' hasn't always been nor always will be experienced as it is today. It is possible, although only barely at the moment, to imagine a time after 'technology', a time when it no longer functions as a grand signifier. All the affects of urgency, speed and relentless dynamism concerning new technologies may well subside, as in the past they have subsided in the wake of technologies such as clocks, railways, electric light, film, television and telephones as they became part of an almost invisible collective infrastructure.

Our current susceptibility to excitement or anxiety about new technologies cannot be completely dismissed as ephemeral, or as the ideological effect of product advertising and commodification. Even if we become more critically aware of the way in which stories told about technology are at odds with technical practices, something more is at stake. Our collectives have been informed by technical mediations in ways that we can scarcely signify. If, like Haraway, we want to learn to remember that we could have been and might yet be different, we need to develop a feel for the peculiar mode of existence of technologies in our collectives. A middle ground between grand narratives of capital 'T' technology and innumerable micro-practices (for instance, the sub-culture of text-messaging on mobile phones) usually regarded as 'cultural' should be mapped. In this middle ground, technical mediations are not the subject, agent or 'motor' of history (Basalla, 1988). This has been an important theme throughout this book. If technology is treated as the subject of history, if it is seen as an agent which forms and shapes cultures, then really there can be no point in trying to understand how we collectively might have been different. We could not have been, except by destroying everything that counts as technology, as the inhabitants of Samuel Butler's *Erewhon* did when they destroyed all machines (Butler, 1921). On the other hand, the frequently voiced feelings of disorientation and inability to keep up with technoscientific change means that we cannot comfortably assert that culture is in control of technology.

The concept of technicity offers a way of thinking about how technical practices are grounded in diverse milieus. It tends to undermine the familiar impasse between technological and human agency because it involves thinking relationally about technical action. It focuses (see Chapter 1) on the concretization of technical elements

and, more importantly for contemporary purposes, on the concretization of technical ensembles composed of such elements. As earlier
chapters have shown, concretization is a complicated transductive
process in which diverse milieus converge in singular zones of intense
interaction. Making a brick, it seems trivial to say, is a transductive
event. The event in which clay and mould exchange properties is a
point of inflexion, not simply a point of intersection or collision. What
counts as the form of a brick and what counts as its matter are effects
of this interaction. They are not its pre-existing constituents. Only
because the habits of thinking of matter and form as separate is so
strong can such an occurrence appear trivial. Moreover, as I have
suggested throughout this book, it does not always occur so transiently
as it does when clay and mould are brought together. The point of
inflexion, or the moment when actors exchange properties, can have
an extended duration. It can involve ongoing interaction, stabilization
and destabilization between different realities. As Stelarc's *Ping Body*,
the genomic databases, or the computer game Avara show, part of the
problem resides in how to delineate and delimit the extent of these
events when it comes to large, distributed networks of technical
elements. The technical ensembles discussed in this book have been
selected because they mark persistent troublespots for classificatory
schemas form/matter, human/non-human or social/technical.

READING TECHNOLOGICAL OBJECTS

At the start of this book, a suggestion was made about reading
technological practices and objects. It was, I said, necessary to learn to
read again because many responses to contemporary technical ensembles swing between repudiation and over-identification. That oscillation prevents us from looking for the emergence of new capacities,
or remembering how we might have been and could be different. The
broad question I started with was: 'how can we acknowledge the
powerful global extension of modern technology, with all its dislocating effects of speed, and yet remain responsive to the specific material
practices and situations in which collectives of humans and non-humans
are involved?' Now we can ask: 'in what ways does the concept of
transduction imply a way of reading technology outside the oscillation

between repudiation and over-identification? How could that opening question now be read differently'? Bruno Latour (1996, 207) enjoins us to read technological mechanisms directly:

> Beyond our infinite respect for the deciphering of Scripture, we need to have infinite respect for the deciphering of *inscriptions*. To propose the description of a technological mechanism is to extract from it precisely the *script that the engineers had transcribed in the mechanisms and automatisms of humans and nonhumans.* (Original author's italics)

Such a reading is possible because various traces and semiotic systems – diagrams, system descriptions, codes and standards, displays, manuals, instructions, warnings, indicators and controls – cluster around technical artefacts and ensembles. Reading and writing form a large part of the operation of contemporary machines. Different codes, protocols and conventions compose much of the infrastructure involved in computation, for instance. These layers of inscription mark the diverse negotiations which a technical object embodies.

The quasi-allegorical reading of the nuclear bomb and Acheulian hand-axe (Chapter 2) was in effect a de-scription of two limit cases, or two practices. The point of this allegory can be put simply: sometimes it is hard to identify the engineers. Both artefacts mark a kind of discursive limit. The hand-axe figures as a threshold for human collectives. Without something like hand-held implements, it is hard to imagine that a collective is human. The nuclear bomb is also troublesome to represent. Toying with such a material-semiotic actor, a political collective wields excessive force. In both cases, the limits of human collectives are in question. One way to read such limits is to follow the lead of existing work in science and technology studies. As we saw, an extensive and rich set of exceptions to the dominant oppositions between the technical and the social have been mapped out in recent studies.

Over the last two decades, science and technology studies (STS) have wrought a sea change in humanities and social science approaches to technology. Its accounts of the entwining of human and non-human agency deeply affect much of this book. From amid its case studies of particular technical systems, a general pattern emerges in which human and non-human actors, and social and technical factors are linked together inextricably. The vital insight on which I have been drawing concerns the constitution of human collectives as combinations of the living and non-living. For science and technology studies, there is nothing purely human, only mixtures of human and non-human elements joined together by networks of association. As the case of the hand-axe showed, this synthesis through 'mixing' is extremely intimate. It reaches at least as far as the level of neuromuscular organization. However, this book has diverged from that body of work in not relying solely on the idea of *mixing* to account for technical practice.

TECHNICITY AND MEANING

To see the question of technical practices as something more than a mixing of humans and non-humans, we could reconsider the significance of some of the other cases discussed in earlier chapters. In every case, starting from the crude examples of a brick or hand-axe, moving on to clocks, communication networks or computer games, and ranging out to the technical ensembles distributed around the nuclear bomb or the genomics databases, technical practices impinge upon representation and thought in surprising ways. For instance, the case of the brick seems to provide a paradigmatic instance of the distinction between matter and form. It appears to be clay moulded into a regular geometrical shape. That distinction between form and matter not only permeates accounts of embodiment (see Chapter 1), it turns up in different guises in many conceptual frameworks. If, as I have argued, the informing of matter and form can be understood as a transductive event, we are also offered a way of thinking against the grain of most existing accounts of embodiment. In other words, thinking through technical practices affects the way that we think about other problems, such as embodiment, which are usually taken to be distinct from the technical. It is possible to go one step further and, oriented by the quasi-concept of originary technicity, argue that technical practices always touch on the conditions of signification in some way. As the Introduction suggested, originary technicity is the idea that there is no way we can conceptualize or signify the human without already relying on a notion of the technical. Although I have not explored the implications of this idea, it might also mean we that cannot *think* without being in some way technical.

The broad view that within our collectives, technical mediations are subordinate to existing meaning-structures therefore needs reappraisal from the transductive standpoint I am advocating. The emphasis on subordinating technology to signification, especially to linguistic signification, and the goal of symbolizing technology within culture does not take into account the ways in which technical mediations might resist signification. Technical mediations can be thoroughly 'metabolized' in collective life, to the extent that they become an invisible infrastructure, or they can exist at the very limits of the collective. Both possibilities mean that technology cannot be simply represented as such. Stelarc's *Ping Body*, for instance, slows down and renders perceptible the temporal and spatial disjuctions associated with networked information. By rendering visible habituated patterns of anticipation and delay, *Ping Body* shows how something that is normally seen as merely technical (the existence of delays) is also entwined with embodiment. Similarly, an example of how the technical is involved at the limits of signification was introduced in the context of the discussion of a real-time networked computer game (in Chapter 5). I suggested there that the kind of collective information, even only fleetingly, during certain kinds of mediated interaction was difficult to signify using existing notions of community, culture or society. That singular organization of spatial and temporal patterns could not be understood apart from the technical artifacing involved. The difficulty of signifying these kinds of collective formation stems from the fact that they occur at a level of embodied, temporalizing technical action which is usually regarded as prior to, or outside of, signifying processes. Importantly, then, the task of reading technology differently entails reading across technical and cultural domains. It requires different mappings of how embodied practices are temporally and topologically organized.

A wide spectrum of work in the humanities addresses technology through cultural representations in isolation from technical practices. From the transductive standpoint, the risk is that the cultural and historical studies of technology, with their concentration on semiotic and narrative analyses of technologies as texts, might overlook the originary technicity of culture. In this area, Donna Haraway's work is one of the most concerted attempts to go beyond semiotic analyses with its efforts to understand how 'technology . . . turns body into story and vice versa' (Haraway, 1997, 179). If this book has not accessed technical mediations at the level of their cultural representation in literature, film or other media, this does not mean that narrative, textual and semiotic analyses of technologies are not crucially important. For instance, the many comments about speed to be found scattered throughout the book are responses to a particularly prevalent narrative about modern technology. However, I have tried to suspend the assumption that existing languages, narratives and sign systems fully capture all the exchanges, mutations and events within a collective.

POINTS OF SINGULARITY

How then, can we read technology differently? The guiding principle here has been to focus on the points of inflection or singularities within technical ensembles. At these points, actors – bodies, machines, signs, milieus – are catalysed, to echo Rabinow, into new assemblages. In any technical mediation, there are points at which humans and things exchange properties, or at which body becomes story and vice versa. Such points are crucial because they delineate a border between what counts as human and what counts as non-human, between what functions technically and what presents itself as social; they decide finally between what lives and what does not live. These points take different forms. Often they involve a topological twist or a temporal disjunction. Again, we have seen examples which include the escapement of the pendulum clock, the binary split partitioning trees of the real-time animated computer game, the folding of genomic or proteomic sequence data ready for searching and comparison, and the anticipation and delay involved in a gesture with a hand tool or Stelarc's Third Arm. These sites of articulation are not always highly visible, on the surface, or at the interface between technical and non-technical. But on either side of them, more recognizable human and technical figures appear. In looking for these points of articulation amid the complicated ensembles of signs and things presented by contemporary technology, I am not saying that technical mediations are the underlying reality that social conventions over-code for their own purposes. The points of inflection show that technical operations can be analysed as the articulation of diverse realities on each other. Another name for what occurs around these points process might be *eventalization* (Foucault, 1991).

As a consequence of this general approach to technical practices, the problem of the speed of technological change takes on a different complexion. Stories about the accelerating pace of technological innovation and the increasing speed of technical ensembles have abounded for at least a 150 years. These stories contribute to the oscillation between over-identification and repudiation that I have been analysing. They narrate technology as irreversible, monolithic, autonomous and accelerating. Of course, the experience of speed cannot be denied. But concepts of self-present subjectivity usually associated with modernity (e.g. the Cartesian subject) cannot account for the experience of speed associated with technologization except as a loss of orientation (Stiegler, 1996). By contrast, the reading of speed under development in earlier chapters has suggested that the experience of speed is an effect of differences. The point is obvious at one level: there can be no experience of speed except as a change of speed. Here, I am arguing something more. The transductions we have discussed occur at the limits of the collective, at points of inflection where interior becomes exterior, and vice versa. Differences in speed are the consequences of an ongoing process of folding inwardly and outwardly. The efficacy and speed of the Acheulian hand-axe as a projection of human gestures is accompanied, at least approximately, by an intricate neuronal folding. The annihilating energies released by a nuclear detonation ride on a dense flux of calculation enclosed in supercomputer simulations. Transductions, as I have said, are not necessarily instantaneous or spatially encapsulated. We can expect a lack of synchronization between what is disembodied and what is embodied, between what runs ahead and what remains in delay. In terms of both corporealization and temporalization, these dynamics of anticipation and delay mean that technical mediations are not simply present. Each time that a transductive process comes into play, the embodiments and temporalities available within that collective shift, bringing them into tension with existing codings of bodies, places and time. We saw these tensions at work around clocktime, information networks, real-time communication and bioinformatics.

Constant technological innovation makes it difficult to decide whether a particular innovation is significant or new. (Currently, for instance, people are debating whether biotechnological transgenic organisms are totally new or nothing new at all.) For two reasons, I am not sure how important it is to say whether some technical phenomenon is really new or not. First, even if we do find something new or novel, the significance of that judgement could already be somewhat undermined by what Jean-François Lyotard formulates as technology's absorption of an ever-increasing rate of contingency – the idea that technological systems act as increasingly complex buffers, processing social, economic, ecological or natural events (see Chapter 1). There is the possibility that paroxysmal change can be regarded impassively or with indifference. This is a magnified version of the 'scroll blindness' familiar to computer workers who spend a long time scrolling through screens of text. Second, highlighting a single innovation usually avoids the more significant problem of how to recognize the technicity of the ensembles in which our collectives are entwined. As a way of thinking around this impasse, I see more promise in the idea that the experience of disorientation or impossibility of deciding whether something is new or not is also an opportunity to render intelligible something about a collective individuation. Reading technical practices in terms of their production of speed again involves mapping how the effects of speed are produced across an ensemble. It is possible to do this through works that figure this ensemble, such as Stelarc's *Ping Body*, or by tracing how a common technical practice like clocktime is propagated through technical ensembles. The emphasis on the *ensemble* in this book is significant. Both the flash of nuclear detonation and the potentially 'accelerated evolution' attributed to biotechnologies rely on largely invisible infrastructures (e.g. supercomputers, databases, etc.) whose technicity is difficult to represent because it is composed of many different technical elements.

TECHNOLOGIES ARE COLLECTIVE INDIVIDUATIONS

In the transductive reading of technical practices, technologies are collective individuations. The implications of this point are not easy to grasp. Why do some technical ensembles provoke a strong cultural reaction while others remain largely invisible? Why are some objects (gadgets, machines, systems) marked as explicitly technical and others are not? In terms of originary technicity, there is no practice, no signification, no affect for that matter, that does not have something technical about it. Eating, praying, dressing and talking can be seen as having technical facets. Would we want to say, then, that all these practices are also collective individuations? More often than not, such questions have been answered in terms of science and modernity. Technological practices are understood as flowing from scientific knowledge, and as possessing a rational foundation and a potentially colonizing universality that other cultural practices do not. I have drawn on some different theoretical approaches to realign this perception. The work of Heidegger, Haraway and Latour all question the primacy of reason as the foundation of technology. By quite different paths, all of them attribute a deeper temporality and a less homogeneous extension to technology than this perception suggests. Heidegger (see Chapters 3 and 4) understands technology as an event in the history of Being which inscribes how things come to be or appear as they are. That understanding keeps alive the question of who 'we', collectively, are. Latour (drawing on Michel Serres) espouses something similar when he refigures the split between modern and premodern cultures with an account of how collectives are stabilized through translations (inscriptions, delegations, detours) into networks of things. Finally, Haraway, as she maps how troublesome boundary objects such as databases, foetuses or stem cells permit bodies to become story and vice versa, indicates how what counts as technology is a highly site-specific nexus of stories, investments, knowledges and inscriptions, not an a priori category.

The supercomputer, Stelarc's *Ping Body*, the GPS, the computer game and the genomic databases are not randomly chosen examples. They are eventalizations in the sense that Rabinow sketches. Clocktime oscillations, the commodified real-time image, and bioinformatics all materialize at specific junctures where temporal and topological reorganizations of collective life occur. Important dimensions of these eventalizations are *informatic*. Today, we find it difficult to broach the question of technology apart from its informatic dimensions. What does a transductive reading say about information? Viewed transductively, information has close links to corporealization and temporalization. Rather than apprehending information as foreign to living bodies, it suggests that we should attend to the material interfaces and technologies that make disembodiment such a powerful illusion (Hayles, 1999, 47). A parallel approach could be taken to temporality and information: the experience of a loss of time can be met by attention to the processes of timing and synchronization that render real time and acceleration persuasive and encompassing. While 'information' is often understood as a kind of abstract foundation of contemporary technologies (including, as we have seen, biotechnology), it can also be read as a complex of intersecting norms, institutions and scripts plied through technical mediations. The abstract concept of information in reality only has value as a concrete, point by point interlacing of different realities. Simondon's understanding of information as in-formation (see Chapter 1) makes this point directly. Concepts of radical contingency or finitude have magnetized social and political theory for some time now. Radical contingency is an abstract way of saying that collectives cannot expect to find ultimate foundations for their own mode of existence. Judith Butler's theory of citational materiality, Michel Foucault's explorations of the place of the arbitrary, contingent and singular within what is given as universal and necessary, and the theme of originary technicity found in Derrida's work are broad symptoms of this turn to radical contingency. However, technology has not been affirmed in the context of radical contingency. In general terms, it has figured as a delocalizing and homogenizing force, absorbing and neutralizing contingency. It tends to be seen as something that can only veil the abyssal absence of foundation. Consequently, technology is often framed as neutralizing contingency by increasing technical mediation. We think, sometimes hopefully, sometimes anxiously, that machines will obviate the unexpected and that, through them, even at the cost of constant innovation, we can amortize ourselves against the future. This dimension of technical practices cannot be lightly dismissed. Yet, nor can another possibility be dismissed: that technical practices do more than confirm or corrode existing limits of human culture, subjectivity and experience.

References

Agamben, Giorgio (1993a) *The Coming Community*, tr. by Michael Hardt (Minnesota University Press, Minneapolis).

Agamben, Giorgio (1993b) *Infancy and History: The Destruction of Experience*, tr. by Liz Heron (Verso, London).

Agamben, Giorgio (1998) *Homo Sacer: Sovereign Power and Bare Life*, tr. by David Heller-Roazen (Stanford University Press, Stanford).

Ansell Pearson, Keith (1997) *Viroid Life: Perspectives on the Transhuman Condition* (Routledge, London).

Aristotle 1996 *Physics*, tr. by Robin Waterfield (Oxford University Press, Oxford).

Atlan, Henri and Keppel, Moshe (1990) The cellular computer DNA: program or data. *Bulletin of Mathematical Biology*, 52(3): 335–48.

Balsamo, Anne (1996) *Technologies of the Gendered Body: Reading Cyborg Women* (Duke University Press, Durham).

Basalla, George (1988) *The Evolution of Technology* (Cambridge University Press, Cambridge).

Beardsley, Tim (1996) Vital data. *Scientific American*, 274(3): 76–81.

Beardsworth, Richard (1995) From a genealogy of matter to a politics of memory: Stiegler's thinking of technics. *Tekhnema: Journal of Philosophy and Technology*, 1(22): 85–114.

Beardsworth, Richard (1998) Thinking technicity. *Cultural Values*, 2(1): 70–86.

Bennington, Geoffrey (1996) Emergencies. *Oxford Literary Review*, 6(1): 76–216.

Berg, Paul and Singer, Maxine (1992) *Dealing with Genes: The Language of Heredity* (University Science Books, Mill Valley, CA).

Boguski, Mark S. (1999) Biosequence exegesis. *Science*, 286 (15 October), 453–5.

Bolter, Jay David and Grusin, Richard (1999) *Remediation: Understanding New Media* (MIT Press, Cambridge, MA).

Borst, Arno (1993) *The Ordering of Time: From the Ancient Computus to the Modern Computer* (Polity Press, Cambridge).

Bowker, Geoffrey (1995) Second nature once removed. Time, space and representations. *Time & Society*, 4(1): 47–66.

Brand, Stewart (1999) *The Clock of the Long Now: Time and Responsibility* (Basic Books, New York).

Butler, Judith (1993) *Bodies That Matter: On the Discursive Limits of Sex* (Routledge, London).

Butler, Samuel (1921) *Erewhon* (Jonathan Cape, London).

Calvin, William (1993) The unitary hypothesis: a common neural circuitry for novel manipulations, language, plan-ahead and thowing? In *Tools, Language and Cognition in Human Evolution*, ed. by K.R. Gibson and T. Ingold (Cambridge University Press, Cambridge).

Campbell-Kelly, Martin and Aspray, William (1996) *Computer: A History of the Information Machine* (Basic Books, New York).

Canguilhem, Georges (1994) The concept of life. In *A Vital Rationalist: Selected Writings from Georges Canguilhem*, ed. by Francois Delaporte, tr. by Arthur Goldhammer (Zone Books, New York), 301–20.

Canguilhem, Georges (1991) *The Normal and the Pathological*, tr. by Carolyn R. Fawcett and Robert S. Cohen (Zone Books, New York).

Cantor Charles (1994) Can computational science keep up with evolving technology for genome mapping and sequencing? In *Computational Methods in Genome Research*, ed. by Sándor Suhai (Plenum Press, New York), 1–18.

Cheagh, Pheng (1996) Mattering. *diacritics*, 26(1): 124.

Collins, Francis S., Patrinos, Ari, Jordan, Elke, Chakravarti, Aravinda, Gesteland, Raymond and Walters, LeRoy (1988) New goals for the U.S. Human Genome Project: 1998–2003. *Science*, 282: 682–9.

Combes, Muriel (1999) *Simondon Individue et Collectivité* (PUF, Paris).

Cook, Jeffrey, Zebington, Gary and De Silva, Sam (1997) *Metabody: From Cyborg to Symborg* (CD-ROM, Merlin Integrated Media, Sydney).

Critchley, Simon (1999) *Ethics, Politics, Subjectivity: Essays on Derrida, Levinas and Contemporary French Thought* (Verso Books, London).

Davidson, Ian and Noble, William (1993) Tools and language in human evolution. In *Tools, Language and Cognition in Human Evolution*, ed. by K.R. Gibson and T. Ingold (Cambridge University Press, Cambridge).

Debord, Guy (1995) *The Society of the Spectacle*, tr. by Donald Nicholson-Smith (Zone Books, New York).

Deleuze, Gilles (1993) *The Fold: Leibniz and the Baroque*, tr. by Tom Conley (University of Minnesota Press, Minneapolis).

Deleuze, Gilles (1994) *Difference and Repetition*, tr. by Paul Patton (Columbia University Press, New York).

Derrida, Jacques (1973) *Speech and Phenomena and Other Essays on Husserl's Theory of Signs*, tr. by David B. Allison (Northwestern University Press, Evanston).

Derrida, Jacques (1984) No apocalypse, not now (full speed ahead, seven missiles, seven missives). *diacritics*, tr. by Catherine Porter (summer, 1984), 20–31.

Derrida, Jacques 1989 *Of Spirit: Heidegger and the Question*, tr. by Geoffrey Bennington and Rachel Bowlby (University of Chicago Press, Chicago).

Derrida, Jacques (1993) The rhetoric of drugs. An interview, tr. by Michael Israel. *Differences: A Journal Of Feminist Cultural Studies*, 5.1: 1–10.

Derrida, Jacques (1994) *Specters of Marx: The State of the Debt, the Work of Mourning and the New International*, tr. by Peggy Kamuf (Routledge, New York).

Derrida, Jacques and Stiegler, Bernard (1996) *Échographies de la télévision* (Galilée-INA, Paris).

Doolittle, Russell F. (1990) What we have learned and will learn from sequence databases. *Computers and DNA. Santa Fe Institute Studies in the Science of Complexity*, vol 7, ed. by George I. Bell and Thoman G. Marr (Addison-Wesley, New York), 21–31.

Dreyfus, Hubert (1995) Heidegger on gaining a free relation to technology. In *Technology and the Politics of Knowledge*, ed. by Andrew Feenberg and Alastair Hannay (Indiana University Press, Bloomington), 97–107.

Dumouchel, Paul (1995) Gilbert Simondon's plea for a philosophy of technology. In *Technology and the Politics of Knowledge*, ed. by Andrew Feenberg and Alastair Hannay (Indiana University Press, Bloomington).

Edwards, Paul N. (1996) *The Closed World: Computers and the Politics of Discourse in Cold War America* (MIT Press, Cambridge, MA).

Elias, Norbert (1993) *Time: An Essay* (Blackwell, Oxford).

Evans, William J. and Relling, Mary V. (1999) Pharmacogenomics: translating functional genomics into rational therapeutics. *Science*, 286 (15 October), 487–91.

Feenberg, Andrew (1991) *Critical Theory of Technology* (Oxford University Press, Oxford).

Foucault, Michel (1978) *History of Sexuality, Volume I: An Introduction*, tr. by Robert Hurley (Random House, New York).

Foucault, Michel (1984) What is enlightenment? In *The Foucault Reader*, ed. by Paul Rabinow (Penguin Books, Hammondsworth).

Foucault, Michel (1991) Questions of method. In *The Foucault Effect: Studies in Governmentality*, ed. by G. Burchell, C. Gordon and P. Miller (University of Chicago Press, Chicago), 73–104.

Fowler, Cary (1995) Biotechnology, patents and the Third World. In *Biopolitics: A feminist and ecological reader on biotechnology* (Zed Books Ltd/ Third World Network, London), 214–25.

Fox Keller, Evelyn (1992) *Secrets of Life, Secrets of Death: Essays on Language, Gender and Science* (Routledge, New York).

Fox Keller, Evelyn (1995) *Refiguring Life: Metaphors of Twentieth-Century Biology* (Columbia University Press, New York).

Freeman, Mae and Freeman, Ira (1962) *You Will Go to the Moon* (Random House, New York).

Fuchs, Henry, Kedem, Z. and Naylor, Brian (1980) On visible surface generation by a priori free structures. *Conference Proceedings of SIGGRAPH '80* 14(3): 124–33.

Gale, Mike and Devos, Katrien (1998) Plant comparative genetics after 10 years. *Science*, 282 (23 October), 656.

Gelbart, William M. (1998) Databases in genomic research. *Science*, 282 (23 October), 659–61.

Gilbert, Walter (1992) A vision of the Holy Grail. In *The Code of Codes: Scientific and Social Issues in the Human Genome Project*, ed. by Daniel J. Kevles and Leroy Hood (Harvard University Press, Cambridge, MA).

Grosz, Elizabeth (1994) *Volatile Bodies: Towards a Corporeal Feminism* (Indiana University Press, Bloomington).

Habermas, Jürgen (1987) Technology and science as 'Ideology'. In *Towards a Rational Society* (Polity Press, Cambridge).

Haraway, Donna J. (1997) *Modest_Witness@Second_Millennium. FemaleMan©_ Meets_OncoMouse™ Feminism and Technoscience* (Routledge, New York).

Haseltine, William A. (1997) Discovering genes for new medicines. *Scientific American*, 276 (March), 78–83.

Hayles, Katherine N. (1999) *How We Became Posthuman: Virtual Bodies in Cybernetics, Literature, and Informatics* (University of Chicago Press, Chicago).

Heidegger, Martin (1954) *Vorträge und Aufsätze* (Verlage Gunther Neske, Stuttgart).

Heidegger, Martin (1957) *Identität und Differenz* (Verlag Gunther Neske, Pfullingen).

Heidegger, Martin (1962) *Being and Time*, tr. by John Macquarrie and Edward Robinson (Basil Blackwell, Oxford).

Heidegger, Martin (1972) *On Time and Being* (Harper & Row, New York).

Heidegger, Martin (1977) *The Question Concerning Technology and Other Essays*, tr. by William Lovitt (Harper Torchbooks, New York).

Heidegger, Martin (1994) Einblick in das was ist. In *Bremer and Freiburger Vorträge* (1949) *Gesamtausgabe* Band 79, ed. by Petra Jaeger (Vittorio Klostermann, Frankfurt am Main), 3–77.

Heidegger, Martin and Wisser, Richard (1988) Martin Heidegger im Ges präch mit Richard Wisser. In *Antwort: Martin Heidegger im Gespräch*, ed. by Günther Neske and Emil Kettering (Verlag Günther Neske, Pfullingen).

Hillis, Daniel (1999) Clock Ideas. www.longnow.org/10kclock/clock.htm

Hood, Leroy (1992) Biology and medicine in the twenty-first century. In *The Code of Codes: Scientific and Social Issues in the Human Genome Project*, ed. by Daniel J. Kevles and Leroy Hood (Harvard University Press, Cambridge, MA), 137–63.

Hopper, Ian (2000) IBM rockets past 10 teraflops. *The Australian IT*, 4 July, 46.

Hottois, Gilbert (1993) *Simondon et la philosophie de la culture technique* (De Boeck Universite, Brussels).

Howard, Ken (2000) The bioinformatics goldrush. *Scientific American*, 283(1): 46–51.

Howse, Derek (1980) *Greenwich and the Discovery of Longitude* (Oxford University Press, Oxford).

Hurd, Cuthbert (1985) A note on early Monte Carlo computations and scientific meetings. *Annals of the History of Computing*, 7(2): 141–55.

Huygens, Christiaan (1986) *Christiaan Huygens' the Pendulum Clock, or, Geometrical Demonstrations Concerning the Motion of Pendula as Applied to Clocks* (Iowa State University Press, Ames).

Ingold, Tim and Gibson, Kathleen R. (1993) *Tools, Language and Cognition in Human Evolution* (Cambridge University Press, Cambridge).

Janicaud, Dominique (1998) *Chronos: Pour l'intelligence du partage temporel* (Bernard Grasset, Paris).

Jones, Steve (1994) *The Language of the Genes: Biology, History and the Evolutionary Future* (Flamingo, London).

Judson, Horace Freeland (1992) A history of the science and technology behind gene mapping and sequencing. In *The Code of Codes: Scientific and Social Issues in the Human Genome Project*, ed. by Daniel J. Kevles, and Leroy Hood (Harvard University Press, Cambridge, MA), 37–80.

Kaplan, Elliott D. (1996) *Understanding GPS: Principles and Applications* (Artech House Publishers, Boston).

Kauffman, Stuart A. (2000) Forget in-vitro – now it's 'in silico'. *Scientific American*, 283(1): 50–51.

Kauffman, Stuart A. (1993) *The Origins of Order: Self-organization and Selection in Evolution* (Oxford University Press, Oxford).

Kevles, Daniel J. and Hood, Leroy (1992) *The Code of Codes: Scientific and Social Issues in the Human Genome Project* (Harvard University Press, Cambridge, MA).

Kitcher, Philip (1992) Gene: current usages. In *Keywords in Evolutionary Biology*, ed. by Evelyn Fox Keller and Elizabeth A. Lloyd (Harvard University Press, Cambridge, MA).

Kittler, Friedrich (1996) Farben und/oder Maschinen denken. In *Synthetische Welten. Kunst, Künstlichkeit und Kommunikationsmedien*, ed. by Eckhard Hammel (Verlag Die Blaue Eule, Essen), 119–32.

Laclau, Ernesto (1995) Subject of politics, politics of the subject. *Differences: a Journal of Feminist Cultural Studies*, 7(1): 146–64.

Laclau, Ernesto (1996) Why do empty signifiers matter to politics? In *Emancipations* (Verso Books, London).

Lander, Eric S. (1996) The new genomics: global views of biology. *Science*, 274 (25 October), 536–9.

Lash, Scott (1999) *Another Modernity: A Different Rationality* (Blackwell Publishers, Oxford).

Latour, Bruno (1988) Irreductions. In *The Pasteurization of France* (Harvard University Press, Cambridge, MA).

Latour, Bruno (1987) *Science in Action: How to Follow Scientists and Engineers Through Society* (Harvard University Press, Cambridge, MA).

Latour, Bruno (1993) *We Have Never Been Modern*, tr. by Catherine Porter (Harvester Wheatsheaf Press, London).

Latour, Bruno (1994a) Pragmatogonies: a mythical account of how humans and nonhumans swap properties. *American Behavioural Scientist*, 37(6): 791–808.

Latour, Bruno (1994b) On technical mediation – philosophy, sociology, geneaology. *Common Knowledge*, 3(2): 29–64.

Latour, Bruno (1995) A door must be either open or shut: a little philosophy of techniques. In *Technology and the Politics of Knowledge*, ed. by Andrew Feenberg and Alistair Hannay (Indiana University Press: Bloomington).

Latour, Bruno (1996) *Aramis, or the Love of Technology*, tr. by Catherine Porter (Harvard University Press, Cambridge, MA).

Latour, Bruno (1997) Trains of thought: Piaget, Formalism and the Fifth Dimension. *Common Knowledge*, 6(3): 170–91.

Latour, Bruno (1999) *Pandora's Hope: Essays on the Reality of Science Studies* (Harvard University Press, Cambridge, MA).

Law, John and Callon, Michel (1995) Agency and the hybrid collective. *South Atlantic Quarterly*, 94(2) (Spring), 481.

Lehrach, Hans, Mott, Richard and Zehetner, Guenther (1994) Informatics and experiments for the Human Genome Experiment. In *Computational Methods in Genome Research*, ed. by Sandor Suhai (Plenum Press, New York), 19–24.

Leroi-Gourhan, André (1993) *Gesture and Speech* (MIT Press, Cambridge, MA).

Lewontin, Richard (1992) The dream of the human genome. *New York Review of Books*, 28 May, 31–40.

Lyotard, Jean-François (1991) Time today. In *The Inhuman: Reflections on Time*, tr. by Geoffrey Bennington and Rachel Bowlby (Stanford University Press, Stanford), 58–77.

MacKenzie, Donald (1998) *Knowing Machines: Essays on Technical Change* (MIT Press, Cambridge, MA).

Markus, György (1986) Marx and the problem of technology. *Language & Production*, Appendix II (Reidel, Dordrecht), 146–87.

Marshall, Eliot (1999) Do-it-yourself gene watching. *Science*, 286 (15 October), 444–7.

Marx, Leo (1999) Information technology in historical perspective. In *High Technology and Low-Income Communities: Prospects for the Positive Use of Advanced Information Technology*, ed. by Daniel A. Schön, Bish Sanyal and William Mitchell (MIT Press, Cambridge, MA), 133–48.

Mayr, Otto (1986) *Authority, Liberty and Automatic Machinery in Early Modern Europe* (Johns Hopkins University Press, Baltimore).

Mitcham, Carl (1994) *Thinking Through Technology: The Path Between Engineering and Philosophy* (University of Chicago Press, Chicago).

Nancy, Jean-Luc (1993) War, law, sovereignty – *Techné*. In *Rethinking Technologies*, ed. by Verena Andermatt Conley (Minnesota University Press, Minneapolis), 28–58.

O'Brien, Stephen J. and Menotti-Raymond, Marilyn (1999) The promise of comparative genomics in mammals. *Science*, 286 (15 October), 459.

Pennisi, Elizabeth (1999) Keeping genome databases clean and up to date. *Science*, 286 (15 October), 447–50.

Pickering, Andrew (1995) *The Mangle of Practice: Time, Agency, and Science* (University of Chicago Press, Chicago).

Plant, Sadie (1997) *Zeros + Ones: Digital Women + The New Technoculture* (Doubleday, New York).

Rabinow, Paul (1992) Artificiality and enlightenment: from sociobiology to biosociality. In *Incorporations*, ed. by Jonathan Crary and Sanford Kwinter (Zone Books, New York).

Rabinow, Paul (1999) *French DNA: Trouble in Purgatory* (University of Chicago Press, Chicago).

Resistance is useless. *New Scientist*, 19 February 2000, 2226: 21.

Rogers, Jane (1999) Gels and genomes. *Science*, 286 (15 October), 429.

Serres, Michel (1982) *The Parasite* (Johns Hopkins University Press, Baltimore).

Serres, Michel (1995) *Genesis* (University of Michigan Press, Ann Arbor).

Serres, Michel and Latour, Bruno (1995) *Conversations on Science, Culture and Time* (University of Michigan Press, Ann Arbor).

Setubal, Joao Carlos and Meidanis, Joao (1997) *Introduction to Computational Molecular Biology* (PWS Publishing Company, Boston).

Shiva, Vandana (1995) Biotechnological development and the conservation of diversity. In *Biopolitics: A Feminist and Ecological Reader on Biotechnology*, ed. by Vandana Shiva and Ingunn Moser (Zed Books Ltd/Third World Network, London), 193–213.

Sigaut François, (1994) Technology, *Companion Encyclopaedia of Anthropology*, ed. by Tim Ingold (Routledge, New York) 420–59.

Simondon, Gilbert (1958) Psycho-sociologique de la technicité'. *Bulletin de l'Ecole Pratique de Psychologie et de Pédagogie de Lyon*, 319–50.

Simondon, Gilbert (1989a) *Du mode d'existence des objets techniques* (Editions Aubier-Montaigne, Paris).

Simondon, Gilbert (1989b) *L'individuation psychique et collective* (Editions Aubier-Montaigne, Paris).

Simondon, Gilbert (1992) *L'Individu et sa genèse physico-biologique Incorporations* (eds. Jonathon Crary and Sanford Kwinter) (Zone Books, New York).

Simondon, Gilbert (1995) *L'Individu et sa genèse physico-biologique* (Éditions Jérôme Millon, Grenoble).

Spivak, Gayatri Chakravorty (1996) Scattered speculations on the question of value. In *The Spivak Reader: Selected Works of Gayatri Chakravorty Spivak*, ed. by Donna Landry and Gerald MacLean (Routledge, New York), 107–40.

Spivak, Gayatri Chakravorty and Rooney, Ellen (1994) In a word. In *The Essential Difference*, ed. by Naomi Schor and Elizabeth Weed (Indiana University Press, Bloomington), 151–84.

Star, Susan Leigh and Ruhleder, Karen (1996) Steps toward an ecology of infrastructure: design and access for large information spaces. *Information Systems Research*, March, 7(1): 111–34.

Stelarc (1997) *Metabody: From Cyborg to Symborg*, CD-ROM (eds. Jeffrey Cook, Gary Zebington and Sam De Silva) (Merlin Integrated Media, Sydney).

Stelarc, www.stelarc.va.com.au/

Stengers, Isabelle and Gille, Didier (1997) Time and representation. In *Power and Invention: Situating Science*, tr. by Paul Bains (University of Minnesota, Minneapolis).

Stiegler, Bernard (1993) Questioning technology and time, tr. by Richard Beardsworth. *Tekhnema Journal of Philosophy and Technology* (Fall 1993), 31–44.

Stiegler, Bernard (1994) *La technique et le temps. 1. La faute d'Épiméthée* (Galilée, Paris).

Stiegler, Bernard (1996) *La technique et le temps. 2. La désorientation* (Galilée, Paris).

Stiegler, Bernard (1998) *Technics and Time: The Fault of Epimetheus*, tr. by Richard Beardsworth and George Collins (Stanford University Press, Stanford).

Suzuki, David and Dressel, Holly (2000) *From Naked Ape to Super Species* (Allen & Unwin, Sydney).

Tristram, Claire (1999) Has GPS lost its way? *Technology Review*, 102(4): 70–74.

Virilio, Paul (1993) The third interval: a critical transition. In *Rethinking Technologies*, ed. by Verena Andermatt Conley (Minnesota University Press, Minneapolis), 3–12.

Virilio, Paul (1995a) *The Art of the Motor*, tr. by Julie Rose (University of Minnesota Press, Minneapolis).

Virilio, Paul (1995b) Speed and information: cyberspace alarm! *CTheory*, tr. by Patrice Riemens. www.ctheory.com/default.asp.

Virilio, Paul (1997) *Open Sky*, tr. by Julie Rose (Verso Books, London).

Virilio, Paul (1998) *The Virilio Reader*, ed. by Der Derian (Blackwell Publishers, Oxford).

Watson, J. and Crick, F. (1953) A structure for Deoxyribose Nucleic Acid. *Nature*, 171, 2 April, 753.

Weber, Samuel (1990) The vaulted eye: remarks on knowledge and professionalism. *Yale French Studies*, 77: 44–60.

Winner, Langdon (1977) *Autonomous Technology: Technics-out-of-Control as a Theme in Political Thought* (MIT Press, Cambridge, MA).

Index

acceleration 1, 32, 80, 119
affect 117
Agamben, Giorgio 146–54, 164
Ansell Pearson, Keith 5, 121
atomic clock 88–9

biopolitical 175
bioinformatics 183–4, 192–3, 196, 199
biotechnology 23, 172 6, 191–3, 198, 201–2
body 20, 29, 42–3, 137, 173, 206
  biological 83
  colonization by technology 2
  corporealization 7, 186–7
  disembodiment 31
  images and 147, 161, 163–4
  originary technicity of 5–6
  radical contingency of 30, 86, 206
  sociotechnical hybrid 7
  technicity of 51–2, 84, 85
  whatever body 152, 154, 165, 169–70
Butler, Judith 29, 36–45, 173, 207

Canguilhem, Georges 10, 177, 195, 201–2
clock, see atomic clock; pendulum clock; time
collective 22, 57, 67, 68, 135, 141, 148, 154–5, 209

living and non-living 112, 201, 203–4
mixed constitution 69, 71, 212
topological complexity 84–5, 173
whatever 147
computation 180
contingency 29–34, 39, 54, 134, 206–7, 218
corporeality, see body
corporealization, see body
culture
  relation to technology 9–10
  limits of 57

data structures 163
databases, gene and protein sequence 182, 183, 185, 187, 190, 193–4, 198–9, 201
delay, see speed
Deleuze, Gilles 108, 191
Derrida, Jacques 5–8, 122–3, 140

Elias, Norbert 93–4, 104
events 31–2

folding 70–1, 73, 131
form
  information 45
  modulation of 107
Foucault, Michel 29, 175, 215
Fox Keller, Evelyn 178

games
    computer   22, 146, 149
    see also play
gesture   135–8, 162, 167–8
Global Positioning System   88, 91,
    109–11
Grosz, Elizabeth   29

hand-tools   20, 59–62, 80–3
Haraway, Donna   7, 34, 177, 181,
    182, 186–7, 208
Heidegger, Martin   4, 24, 33,
    time and temporality   92–3, 104
    essence of technology   124–35
heredity   195, 198
history   24, 39–40, 118–19, 135,
    152, 153
Huygens, Christiaan   88, 99
human   43, 207–8
    historical nature   146
    vs prehuman   65
    vs nonhuman   68, 207
hybridization   189
hylomorphism   15, 45–9

image
    animated   156
    perspectival   157
individuation   18, 49–50, 85, 145,
    153, 191
information   30, 158, 172, 217–18
    commodity value   159–60
    cybernetic   50–1
    genetic   178–9, 187, 203
    information theory   178
    life   44, 177–9
    Simondon's notion   49–52, 137
    temporality   152
infrastructure,   *see* technical ensembles
iterability   38–9, 41

language   134
latency tolerance   166–7

Latour, Bruno   7, 67, 70, 78–80, 95,
    124, 207, 211
Leroi-Gourhan, André   82
life   23, 49–50, 173
    as information   181
    mechanism   179
    technical mediation of   171, 200
living/non-living   41, 43, 52–3, 126,
    136, 173–6
logic of the supplement   3, 7–8
Lyotard, Jean-François   31–2, 154,
    158, 159

machine   52–3, 105–6
Mackenzie, Donald
materiality   34–5, 42, 53–4
matter   37–40, 161
    and form   15, 45–6, 212
mediation   46
modernity   31
    break with pre-modern   63–4, 66,
    71
multiplicity,   *see* collective

nuclear weapons   21, 59–62
    detonation   74–5

pendulum clock   87, 90–1, 97–101
play   150, 151, 154, 168–9
politics   43–4
Polymerase Chain Reaction (PCR)
    190, 193
practice
    discursive   36–7

originary technicity,   *see* technicity

quasi-objects   31, 69

Rabinow, Paul   175, 205
real time,   *see* time
representation
    limits of   112–13

satellite navigation   21
Serres, Michel   69–70
signifying processes   35–6, 38, 59
   discursive limits   65, 142, 211
Simondon, Gilbert   10–11, 16–18,
      104, 117, 126, 141, 174, 192
   see technicity, transduction
simulation   59
   nuclear weapons   74–7
spectacle   146–7, 160
speed   1, 21–2, 62–3, 69, 84, 122,
      134, 208, 215
   and collective topography   78
   delay   122–3, 132, 134, 136, 138,
      139, 160–1, 166–7
   Virilio   120–1
Spivak, Gayatri Chakravorty   42, 160,
      206
Stelarc   116–42
Stengers, Isabelle   98–9, 105
Stiegler, Bernard   6, 8–9, 126, 134,
      140
supercomputing   75–7

technical elements   12–13, 70–1, 192
technical mediation   46, 67–8, 69, 79,
      214
   corporealization   188
   incorporation of   82, 142
   mobilization of relations   72, 82
   modulation   98
technical objects   11, 109
technical ensembles   11, 14, 71, 145,
      172, 188, 192, 201, 216
technicity   11–13, 89, 191, 201
   concretization   12–13, 201
   genesis   108
   iterability   14,
   life   8, 174
   margin of indeterminacy   10–11,
      53–4, 68–9, 105–6, 128

originary   3, 5, 8–9, 212, 217
relationality   14–15
thinking   10, 80, 213
technology
   abstraction   xi, 1, 4, 172, 205
   culture   10, 40–1
   determinism   30
   discursive construction   4–5,
   intelligibility of   3, 112, 140, 213
   as signifier   68, 205–6, 209
   technicity of   18–19
time
   calculability   32
   clock-time   21–2, 75–98, 111
   dating   140
   event   141
   real time   151, 160, 168
   temporality   161
   temporalization   9, 93, 133, 168,
      207
   timing   72
transduction   xi, 15–18, 25 n.1,
      45–8, 118, 174–5, 205, 210, 215
   articulation of diverse realities   77,
      83–4, 95–6, 118
   event   96, 205, 210
   individuation   17, 123, 137, 191,
      216–17
   information   45–6
   living/non-living   18, 23, 173
   metastability   17, 76, 103–4, 107,
      111
   ontogenesis   17, 18
   thinking   18
transindividual   117–18, 137, 138,
      141, 145

Virilio, Paul   91, 119–20, 126, 136,
      140

whatever,   *see* body